Power Systems

Electrical power has been the technological foundation of industrial societies for many years. Although the systems designed to provide and apply electrical energy have reached a high degree of maturity, unforeseen problems are constantly encountered, necessitating the design of more efficient and reliable systems based on novel technologies. The book series Power Systems is aimed at providing detailed, accurate and sound technical information about these new developments in electrical power engineering. It includes topics on power generation, storage and transmission as well as electrical machines. The monographs and advanced textbooks in this series address researchers, lecturers, industrial engineers and senior students in electrical engineering.

Power Systems is indexed in Scopus

More information about this series at http://www.springer.com/series/4622

Nagwa F. Ibrahim • Sobhy S. Dessouky

Design and Implementation of Voltage Source Converters in HVDC Systems

Springer

Nagwa F. Ibrahim
Faculty of Technology and Education
Suez University
Suez, Egypt

Sobhy S. Dessouky
Faculty of Engineering
Port Said University
Port Said, Egypt

ISSN 1612-1287 ISSN 1860-4676 (electronic)
Power Systems
ISBN 978-3-030-51660-4 ISBN 978-3-030-51661-1 (eBook)
https://doi.org/10.1007/978-3-030-51661-1

This Springer imprint is published by the registered company Springer Nature Switzerland AG
The registered company address is: Gewerbestrasse 11, 6330 Cham, Switzerland

This book dedicated to:

My mother, sister, teachers, and friends for their love, encouragement, and endless support.
I give all thanks and gratitude to my dear husband Mohammed Eid and to my little daughter Rodyna, wishing for God to protect them.

Nagwa F. Ibrahim

Preface

With recent developments in semiconductors and control equipment, High Voltage Direct Current that uses voltage source converters (VSC-HVDC) has attracted the growing interest of researchers. The use of VSC technology and Pulse Width Modulation (PWM) has a number of potential advantages: short circuit current reduction, rapid and independent control of the active and reactive power, etc. With such highly favorable advantages, VSC-HVDC is definitely going to be a large part of future transmission and distribution systems. HVDC technology based on VSC technology has been an area of growing interest recently because of its suitability in forming a transmission link for transmitting bulk amount of power.

This book deals with the control of VSC-HVDC. The objective of the book is to understand the control structure of the VSC-HVDC system and establish the tuning criteria for the PI controllers of the converter controllers.

A model of a VSC-based HVDC using PWM Technology is developed. A mathematical model of the control system based on the relationships between voltage and current is described for the VSC. A control system is developed combining an inner current loop controller and outer dc voltage controller. The vector control strategy is studied within this book and corresponding dynamic performance under step changes are also examined, system fault, and the different methods to calculate the current reference to control AC fault is investigated in MATLAB/Simulink simulation package. The simulation results verify that the model can fulfill bi-directional power transfers and fast response control and that the system has good steady state performance. The controller parameters tuned according to the developed tuning criteria is found to provide acceptable system performances.

The presented system is implemented using high-speed digital signal processor (DSP1104). The total harmonic distortion (THD) improved from 17.5% to 3.15%. Experimental results that depict the effectiveness of the proposed control method for an inverter are presented.

Suez, Egypt — Nagwa F. Ibrahim
Port Said, Egypt — Sobhy S. Dessouky

Contents

About the Author

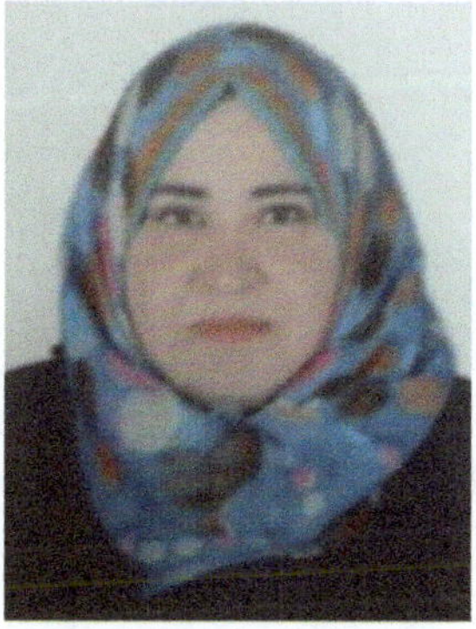

Nagwa F. Ibrahim received her B.S. degree from the Faculty of Industrial Education, Suez Canal University, Suez, Egypt, in 2008; M.Sc. degree from the Faculty of Industrial Education, Suez University, Suez, Egypt, in 2014; and Ph.D. degree from the Faculty of Industrial Education, Suez University, Suez, Egypt, in 2019. She is currently an Assistant Professor with the Department of Electrical Power and Machine, Faculty of Technology and Education, Suez University. Her research interests are in the area of renewable energy sources, power system protection, power electronics, high voltage direct current (HVDC), control of power electronic converters, and electrical machine drives.

Sobhy Serry Dessouky was born in Dakahlie, Egypt, in 1946. He received B.Sc. (1970) and M.Sc. (1977) degrees in Electrical Engineering from Suez Canal University and Helwan University, respectively. Dr. Dessouky received his Ph.D. degree from TU, Dresden, Germany, in 1982. From October 1970 to 1975, he worked in the Faculty of Engineering, Suez Canal University, as Demonstrator. Dr. Dessouky worked as Demonstrator from 1975 to 1977 in the Faculty of Engineering, Helwan University. In 1977, he worked as Assistant Lecturer in the Department of Electrical Engineering, Faculty of Engineering, Suez Canal University. From 1983 to 1987, Dr. Dessouky worked as Assistant Professor (Lecturer) in the Department of Electrical Engineering, Faculty of Engineering, Suez

Canal University, Port Said Campus. In 1987, he was promoted to Associate Professor in the same Department. In 1991, Dr. Dessouky became a Full Professor of Electrical power and H.V. Engineering. He is a member of IEEE since 1996. In parallel, he worked as a Department Chair, Vice Dean for Community Affairs and Environment, and Director of the Engineering Research Center for developing technological planning in Suez Canal University.

List of Abbreviations

2LC	Two-Level Converter
A/D	Analogue to digital converter
AARC	Average Active Reactive Control
AC	Alternating Current
APC	Active-Power Controller
APC-2LC	Active-Power Controlled Two-Level Converter
BPSC	Balance Positive Sequence Control
CBs	Circuit Breakers
CCC	Capacitor Commutated Converter
CSC	Current Source Converter
DC	Direct Current
DPC	Direct Current Control
DSP	Digital Signal Processor
FFT	Fast Fourier Transform
FRT	Fault Ride Through
GTOs	Gate Turn-Off Thyristors
HCC	Hysteresis Current Control
HVAC	High Voltage Alternating Current
HVDC	High Voltage Direct Current
IARC	Instantaneous Active Reactive Control
IGBT	Insulated Gate Bipolar Transistor
LCC	Line Commutated Converter
L-L	Line To Line Fault
MTDC	Multi-terminal HVDC system
NF	Notch Filter
PCC	Point of Common Coupling
PI control	Proportional plus Integral control
PLL	Phase Lock Loop
PNSC	Positive Negative Sequence Control

PWM	Pulse Width Modulation
SCR	Short Circuit Ratio
SL-G	Single Line to Ground fault
THD	Total Harmonic Distortion
VSC	Voltage Source Converter
VSC-HVDC	Voltage Source Converter based High Voltage Direct Current
VSI	Voltage Source Inverter

List of Symbols

a, b, c	The three AC phase components
R_g, L_g	Resistance and inductance of the grid side reactor
R, L	Resistance and inductance of the converter side reacto.
C_f	Capacitance of the input filter
R_d	Damping resistance of the input filter
Z_b, C_b, ω_b	Base impedance, capacitance and pulsation
f_n , ω_n	Grid frequency and pulsation
f_{res}, ω_{res}	Resonance frequency and pulsation of the input filter
f_s	Switching frequency of the converter
i_g, I_g	Grid current and its rms value
v_o, V_o	Dc voltage and its average value
δ_c, τ, δ_v	Ac current and dc voltage constants
$k_{p,c}$, $T_{I,c}$	Current controller gain and time constant
$k_{p,v}$, $T_{I,v}$	Voltage controller gain and time constant
P_n	Average value of the dc power
V_n	Normal voltage
ΔI_L	Current ripple
C_b	Base capacitor
L_g	Grid inductance
X_t	Inductance of transformer
$\overrightarrow{\vec{D}Vsa^{+}}$	Characteristic of voltage sags fault
$\vec{D}$	Phase angel jumping
K_{IP}	Integral control with gains for active power
K_{IQ}	Integral control with gains for reactive power

List of Figures

List of Tables

Chapter 1
Introduction

1.1 Background and Motivation

The use of voltage source converter-based high-voltage direct current (VSC-HVDC) systems is considered to be a major step in facilitating long-distance power transfer and integrating remotely located renewable energy sources to major consumption centers. First introduced in 1997, with the commissioning of a 3 MW technology demonstrator in Sweden [1], VSC technology has improved drastically over the years in terms of power and voltage rating, harmonic performance, and losses [2, 3].

VSC-HVDC is a recent technology, free of several constraints associated with the thyristor-based line commutated converter (LCC) technology, with added degrees of freedom such as independent control of active and reactive power. Additionally, VSC stations can be connected to weak AC grids and even perform black start in contrast to LCC stations that can only be connected to relatively strong AC grids. This also represents a limitation for the LCC-based technology when it comes to integration of renewable power generation units (e.g., wind farms), which usually comprise weak grids due to their low short-circuit ratio (SCR).

Furthermore, the VSC eliminates the need for telecommunication links between stations, which is otherwise a necessity in LCC-HVDC to perform the reversal of power flow. An LCC can reverse its power flow only by reversing the polarity of its direct voltage. When it comes to large-scale multiterminal HVDC (MTDC) systems, where all stations share the same DC link, such a feature prohibits the use of LCC, as there can no longer be independent power direction control at the stations. A common DC-link voltage polarity does not hinder the use of VSC that achieves power flow reversal by reversing the direct current. This property renders the VSC an ideal candidate for implementation in MTDC applications. The introduction of power electronics in power systems has offered a breakthrough in terms of controllability and stability. In turn, this has led to an increased possibility of interactions between the systems components. Two alternative methods are available for the

N. F. Ibrahim, S. S. Dessouky, *Design and Implementation of Voltage Source Converters in HVDC Systems*, Power Systems,
https://doi.org/10.1007/978-3-030-51661-1_1

connection of wind farms to the grid: HVAC and HVDC. For an optimum result, all possible alternatives need to be analyzed, keeping in view the system requirement and the future expansion of the system. Although HVAC provides simplest and economic connection method for short distances, HVDC transmission becomes the only feasible option for connection of a wind farm when the distance exceeds 100–150 km [4]. HVDC traditionally has been used to transfer large amounts of power over long distances. However, for traditional HVDC, the reactive power cannot be controlled independently of the active power.

The HVDC technology based on VSC has recently been an area of growing interest due to a number of factors, like its modularity, independence of AC network, independent control of active and reactive power, low power operation and power reversal, etc. The voltage source technologies also facilitate the connection of several converters to a common DC bus, forming an HVDC grid [5]. VSC-based HVDC systems incorporate the control strategy based on vector control, which is most widely used in the control of three-phase pulse-width modulated (PWM) converters. In PWM converters for AC applications, vector control systems can be utilized to obtain independent control of the active and reactive powers.

1.2 Problem Definition

In PWM converters for AC applications, vector control systems can be utilized to obtain independent control of the active and reactive powers. The characteristics of vector control is that vectors of AC currents and voltages occur as constant vectors in steady state, and hence, static errors in the control system can be avoided by using PI controllers.

The design of controllers is based on two tasks: determining the structure of the controller and adjusting the controller's parameters to give an optimal system performance. The design is normally done with complete knowledge of the system, which is normally described by a linear, time invariant, continuous, or discrete time model. The structure and parameters of the controller are chosen such that the system's response can meet certain qualitative criteria: stable response, appropriate disturbance handling capabilities, robustness, and speed.

The PI controller is the most commonly used control algorithm. PI controllers have traditionally been tuned empirically. There also exist some tuning rules developed to be used for the design of these controllers for industrial applications and electric drives. Design of PI controllers is important because of the significant impact it may give to system response. An efficient design method should be able to provide robust controller parameters. Adequate performance of VSC-HVDC system under diverse operating conditions depends on the selection of robust parameters for the control system.

This book investigates the possible fault ride through (FRT) methods to deal with AC fault. The model and control principle of HVDC converter will be first developed. After that, several FRT methods will be developed and implemented. The system will be tested under short-circuit conditions to test the capability and effectiveness of the proposed FRT methods to handle these situations by different methods to reference currents.

1.3 Objectives

The book is basically focused on the analysis of the control structure. The objectives of the study are mentioned below:

Good understanding of the VSC-based HVDC transmission system concept and of its behavior
Modeling of the VSC-based HVDC transmission system using MATLAB/Simulink
Implementation of control strategies for the VSC-based HVDC transmission system
Analysis of the developed system behavior under different conditions (normal and fault conditions)
Laboratory validation of the developed MATLAB/Simulink model using a small-scale setup and a dSPACE platform

1.4 Book Outline

This book studies the control of voltage source converter-based HVDC power transmission and is structured into six chapters.

The first chapter presents the introduction to the studied subject, containing a short background, the problem formulation, and the objectives of the current book.

The second chapter presents an overview of VSC-based HVDC transmission system. Firstly, a comparison between HVAC and HVDC transmission is presented. The applications and the configuration of HVDC system and its advantages are also discussed.

The third chapter deals with the modeling of the system. Each component is separately explained by presenting its role in the system. The design of the control strategies for the VSC-based HVDC power transmission is investigated. The PLL technique used to synchronize the developed system with the grid voltages is also presented.

The fourth chapter presents different DC and AC fault types, a detailed theoretical analysis of fault, and different control strategies in AC fault.

The fifth chapter presents the design of the current controller, DC voltage controller, active power controllers, different types of DC fault and AC fault, recovery methods for pole-to-pole fault, backup protection of rectifier converter by tripping the grid connection, and different current controllers in AC faults. The tuning process of these controllers was realized using control tool provided by MATLAB/Simulink.

The sixth chapter presents the laboratory implementation of the hysteresis current controller strategy, in order to validate the simulation results.

Finally, the conclusions of the book and ideas for the future work are pointed out in Chap. 7.

Chapter 2
High-Voltage Direct Current Transmission

2.1 Introduction

The HVDC transmission is advantageous for power delivery over long distances and asynchronous interconnections by using overhead lines or underground cables. One of the most important aspects of HVDC systems is its fast and stable controllability [6]. Until recently, the classic HVDC transmission based on thyristors was used for power conversion from AC to DC and vice versa. The appearance of voltage source converter (VSC) makes use of more advanced semiconductor technology instead of thyristors. The VSC-based HVDC installations have several advantages compared to classic HVDC transmission such as independent control of active and reactive power and separate power systems interconnection. VSC-HVDC is also used to reverse the power flow direction without changing the polarity of DC voltage (in multiterminal DC systems), and there are no requirements of fast communication between the two converter stations [7].

2.2 HVDC Converter Arrangements

HVDC converter bridges and cables can be arranged into a number of configurations for effective utilization [8]. Converter bridges may be arranged either monopolar or bipolar as shown in Fig. 2.1.

Monopolar HVDC system is presented in Fig. 2.1a. In this configuration, two converters are used which are separated by a single pole line, and a positive or a negative DC voltage is used [9]. The main properties of monopolar HVDC system are:

N. F. Ibrahim, S. S. Dessouky, *Design and Implementation of Voltage Source Converters in HVDC Systems*, Power Systems,
https://doi.org/10.1007/978-3-030-51661-1_2

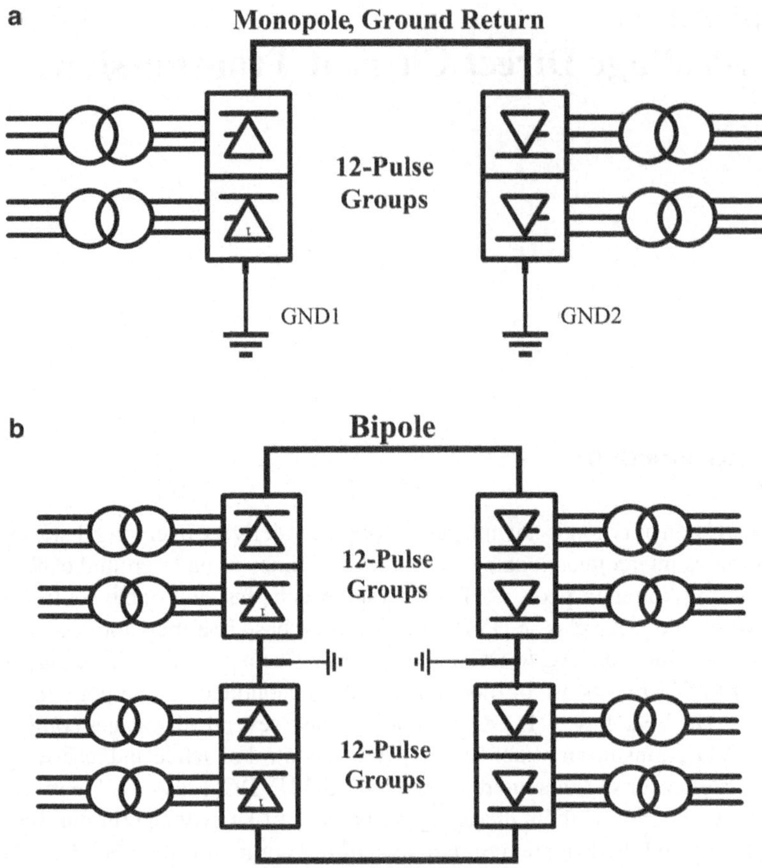

Fig. 2.1 Monopolar (**a**) and bipolar (**b**) connection of HVDC converter bridges

- Single high-voltage conductor.
- Current return through the ground.
- Many existing subsea cable transmissions use monopolar system.

Bipolar HVDC system is presented in Fig. 2.1b. This is the most commonly used configuration of HVDC system in applications where overhead lines are used to transmit power. In fact, the bipolar system consists of two monopolar systems [10]. The bipolar current source converter (CSC) system is presented in Fig. 2.1b. The main properties of bipolar HVDC system are:

- Two conductors, positive and negative polarity.
- Midpoint connected to the ground.
- The connection between the two sets of converters is grounded either at one or at both ends.
- One of the poles can continue to transmit power in case the other one is out of service.

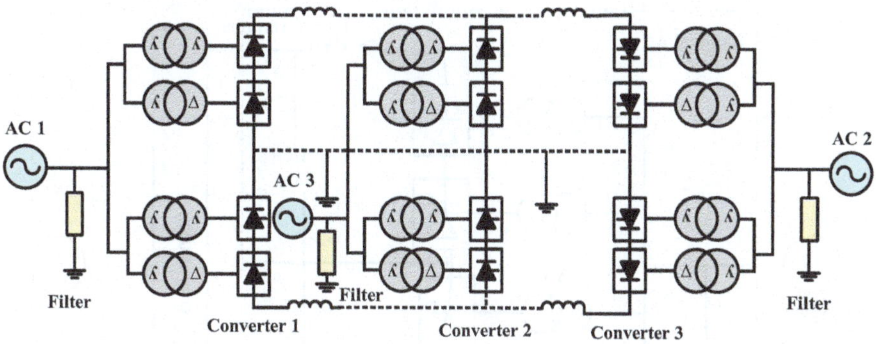

Fig. 2.2 Multiterminal HVDC system [11, 12]

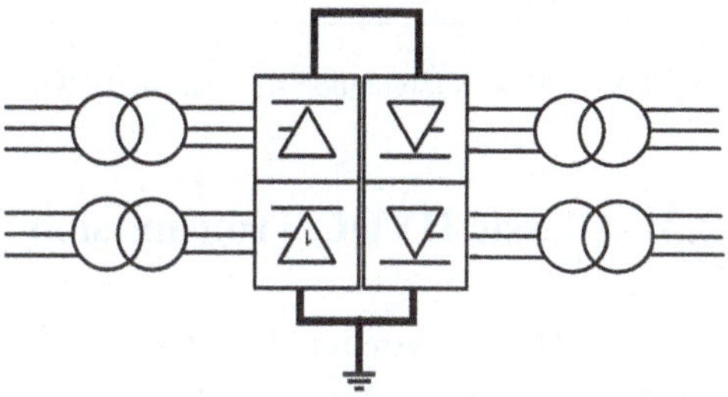

Fig. 2.3 Back-to-back converter system

Multiterminal HVDC system: In the configuration of Fig. 2.2, there are more than two sets of converters like in the bipolar case. Thus, converters one and three can operate as rectifiers while converter two operates as an inverter. Operating in the opposite order, converter two can operate as a rectifier and converters one and three as inverters [11, 12].

Back-to-back HVDC system: In the case of Fig. 2.3, the two converter stations are located at the same site, and no transmission line or cable is required between the converter bridges. The connection may be monopolar or bipolar. The two interconnected AC systems may have the same or different rated frequency, i.e., 50 Hz or 60 Hz. This way of connection can also be used to stabilize the AC system.

The back-to-back converter consists of a rectifier and an inverter connected with a common DC link. The properties of this combination are [12]:

- The DC-link voltage must be higher than the peak line operating voltage.
- The DC-link voltage is regulated by controlling the power flow to the AC grid.
- The possibility of fast control of the power flow.

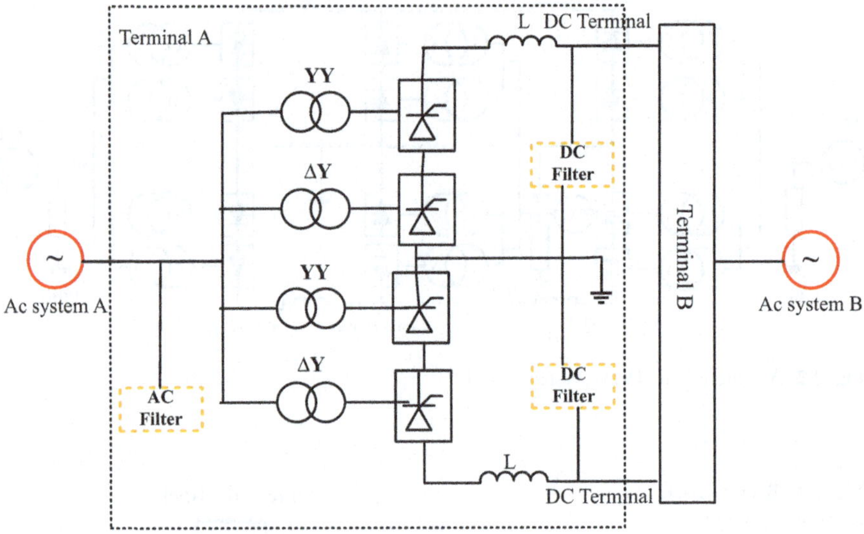

Fig. 2.4 A basic configuration for a classic HVDC system

2.3 Classic HVDC Transmission

A classic HVDC system in Fig. 2.4 consists of AC filters or other reactive compensation equipment, converter transformers, converters, DC reactors, DC filters, and DC lines or cables [13, 14].

The dominant valve type in classic HVDC converter is a thyristor. Thyristors are able to conduct high currents, in a range of several kilo-amps, and can block high voltages [15]. The 6-pulse bridge of Fig. 2.4, as the basic converter unit of classic HVDC, can be used both as a rectifier and as an inverter. A 12-pulse converter bridge can be built by connecting two 6-pulse bridges in parallel or in series. Each bridge consists of defined amount of series-connected thyristors. The bridges are connected separately to the AC system by dint of converter transformers, one with YY winding structure and another with Y-Δ winding structure. In this way, the fifth and seventh harmonic currents through the two transformers are in opposite phase that helps to reduce the distortion in the ac system [16].

The HVDC converters are HVDC system's most vital part. They implement the conversion from AC to DC at rectifier side and from DC to AC at inverter side. HVDC converters are connected to the AC system by dint of converter transformers. As already mentioned before, CSCs are used in classic HVDC transmission. On the AC side of the converter, CSC acts as a constant voltage source. It requires a capacitor as energy storage device, large AC filters for harmonic elimination, and a reactive power supply for power factor correction. On the DC side of the converter, CSC acts as a constant current source. In this case, CSC requires an inductor as its energy storage device and DC filters that provide appropriate fault current-limiting features. The main benefit of CSC is relatively low switching losses.

2.4 VSC-Based HVDC System

VSCs utilize self-commutating switches, e.g., gate turn-off thyristors (GTOs) or insulated gate bipolar transistors (IGBTs), which can be turned on or off in a controlled manner. VSCs operate at high switching frequency 10KHZ utilizing (PWM) technique.

Generally, the new transmission technology has the following advantages compared to classic thyristor-based HVDC [17]:

- Possibility to control the reactive power independent of the active power (to or from the converter) without any needs for extra compensating equipment
- Little risk of commutation failures in the converter
- Possibility to connect the VSC-HVDC system to a "weak" AC grid or even to one where no generation source is available and naturally the short-circuit level is very low [18]
- Faster dynamic response due to higher switching frequency operation (phase-controlled), which further results in reduced need for filtering and hence smaller filter size [19]
- Minimal environmental impact

However, VSC transmission has some disadvantages, which include potentially high switching power losses and high capital costs compared to classic HVDC, though the technology continues to evolve (Table 2.1).

2.4.1 Configuration of VSC-HVDC

The two VSCs may be seen as the core of VSC-HVDC transmission system topology. One of the VSCs operates as a rectifier while the other one as an inverter. Both converters utilize IGBT power semiconductors as switches. The two VSC stations are connected via a DC transmission line. Mainly, two basic configurations of VSCs are used in HVDC transmission system [20]. These are the two-level VSC converter, presented in Fig. 2.5a, and the three-level VSC converter, which is

Table 2.1 Comparison of classic and VSC-HVDC systems [4]

Function	Classic HVDC	VSC- HVDC
Converter valves	Thyristor	IGBT
Connection valve-AC grid	Converter transformer	Series reactor (+ transformer)
Filtering and reactive compensation	50% in filters and shunt capacitors	Only small filter
DC current smoothing	Smoothing reactor + DC filter	DC capacitor
Telecom between converter station controls	Needed	Not needed

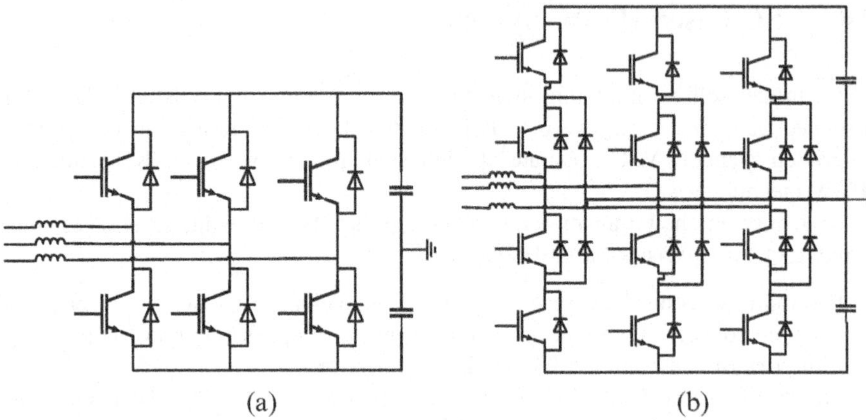

Fig. 2.5 The topology of the VSC: (**a**) two-level VSC and (**b**) three-level VSC

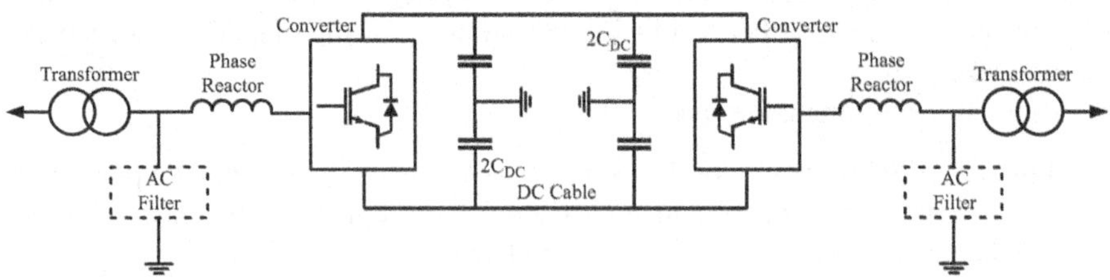

Fig. 2.6 Typical VSC-HVDC system

presented in Fig. 2.5b. Typical configuration of VSC-based HVDC transmission system is presented in Fig. 2.6.

VSC-based HVDC transmission system consists of two VSCs, transformers, phase reactors, AC filters, DC-link capacitors, and DC cables [21]. The sending end and receiving end of VSC-HVDC have the similar configuration, one operating as a rectifier and another as an inverter. In this work, the two converters are connected back to back. Normally, converters are connected to the AC system by means of transformers. The most important function of transformers is to transform the AC voltage level to the DC voltage level. Usually, they are single-phase three-winding type, but depending on the transportation requirements and the rated power, they can be arranged in other ways [22].

The phase reactors are used for controlling active and reactive power flow by regulating currents through them. The reactors also serve as AC filters and therefore reduce the high-frequency harmonic contents of the AC currents which are caused by the switching operation of the IGBTs. The DC side contains two equally sized capacitors. The size of these capacitors depends on the required DC voltage. The primary objective of the DC capacitor is to provide a low inductive path for the turned-off current [23] and energy storage to be able to control the power flow. The capacitor also reduces the voltage ripple on the DC side. AC filters prevent the voltage harmonics entering the AC system. New type of DC cables is used in

VSC-HVDC applications, where the insulation is made of an extruded polymer that is particularly resistant to DC voltage. Polymeric cables are the preferred choice for HVDC, mainly because of their mechanical strength, flexibility, and low weight.

VSC on the AC side acts as a constant current source and therefore requires an inductor as its energy storage device. A small AC filter for harmonic elimination is also required on the AC side. On the DC side, VSC acts as a constant voltage source, and it requires a capacitor as its energy storage device. Energy storage capacitor here provides DC filtering capability. When compared to classic HVDC that are based on CSC, VSC-based HVDC has relatively high switching losses, but by using a soft-switching commutation scheme, the switching losses are considerably reduced [24].

The transfer of power in the case of VSC-based HVDC transmission systems is controlled in the same way as in the case of classic HVDC transmission. The inverter side controls the active power, while the rectifier side controls the DC voltage [25]. The control system of the VSC-based HVDC is realized by using a fast inner current control loop and several outer control loops, depending on the application [26].

The control system of VSC-HVDC has at its base level a fast inner current control loop that controls the AC currents. The AC current references are provided by the outer controllers [27]. The slower outer controllers include the DC voltage controller, the AC voltage controller, the active power controller, the reactive power controller, and the frequency controller. Thus, the reference of the active current can be obtained from the DC voltage controller, from the active power controller, or from the frequency controller. On the other hand, the reference of the reactive current can be derived from the reactive power controller or from the AC voltage controller [28].

The most relevant components of the HVDC system are the following:

- The thyristor or IGBT valves that realize the conversion from AC to DC and thus are the main components of any HVDC converter
- The converter transformers that transform the voltage level of the AC busbar to the required entry voltage level of the converter
- The smoothing reactor, which has the main functions of prevention of the discontinuous current, limitation of the DC fault currents, and prevention of resonance in the DC circuits
- The AC harmonic filters, which absorb harmonic currents generated by the HVDC converter and supply reactive power
- DC transmission circuit consisting of DC transmission line, cable, DC switches, and earth electrode

2.4.2 Operation of VSC-HVDC

The VSC-HVDC system is based on self-commutated PWM technique. PWM generates pulse-width modulated signal by comparing the instantaneous magnitude of a triangular waveform with sinusoidal input reference. Thus, the VSC can produce

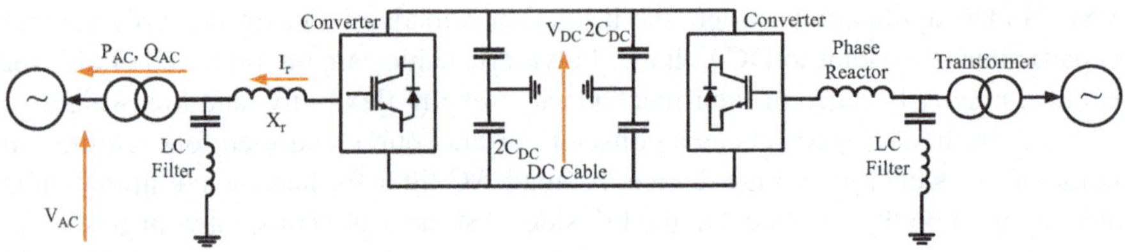

Fig. 2.7 Single-line diagram for VSC-HVDC link

Fig. 2.8 Typical P-Q diagram of VSC-HVDC transmission [30]

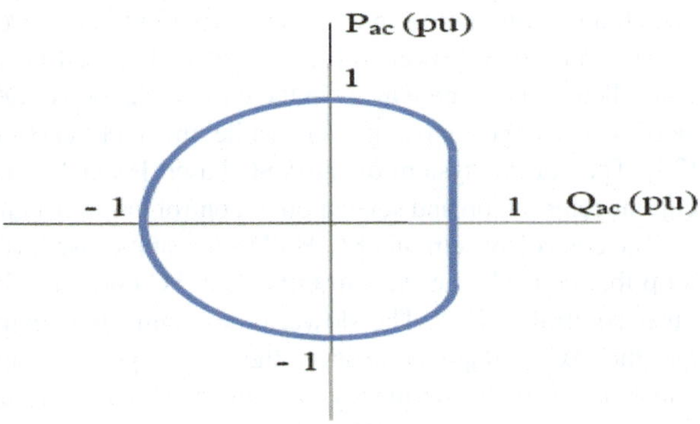

its own voltage waveform independent of the AC system. PWM controls the average output voltage during a very short period, which is called switching period. PWM technique uses a sinusoidal reference signal and produces the desired output voltage. The independent control of active and reactive power can be achieved by means of PWM template that can be changed almost instantly. Changing the fundamental frequency voltage phase angle across the series reactor controls the power, whereas changing the fundamental frequency voltage magnitude across the series reactor controls the reactive power [29]. Two VSC converters are connected together in back-to back configuration forming a transmission link as shown in Fig. 2.7.

By implementing the PWM technique, each converter can be controlled in all four quadrants of Fig. 2.8, while the constant DC voltage is maintained. Reactive power on every terminal can be controlled by regulating AC voltage independently of transmission power level or voltage rating of the converters [30].

2.4.3 *Difference from Classical HVDC and Advantages*

The main difference in operation between classic HVDC and VSC-HVDC is the higher controllability of the latter. This leads to a number of new advantages and applications, some of which are given below [31]:

- Classic HVDC terminals can provide limited control of reactive power by means of switching of filters and shunt banks and to some level by firing angle control. However, this control requires additional equipment and therefore extra cost.

With PWM, VSC-HVDC offers the possibility to control both active and reactive power independently. Hence, independent control of active and reactive power is possible without any needs for extra compensating equipment.

- The reactive power capabilities of VSC-HVDC can be used to control the AC network voltages and thereby contribute to an enhanced power quality. Furthermore, the faster response due to increased switching frequency (PWM) offers new levels of performance regarding power quality control such as flicker and mitigation of voltage dips, harmonics, etc.
- HVDC light does not rely on the AC network's ability to keep the voltage and frequency stable. Independence of AC network makes it less sensitive for disturbances in the AC network, and AC faults do not drastically affect the DC side. If AC systems have ground faults or short circuits, whereupon the AC voltage drops, the DC power transmitted is automatically reduced to a predetermined value. Hence, there is no contribution to the short-circuit currents.
- By controlling grid voltage level, VSC can reduce losses in the connected grid.
- Disturbances in the AC system may lead to commutation failures in classic HVDC system. As VSC-HVDC uses self-commutating semiconductor devices, the risk of commutation failures is significantly reduced.
- The control systems on rectifier and inverter side operate independently of each other. They do not depend on a telecommunication connection. This improves the speed and the reliability of the controller.
- The VSC converter is able to create its own AC voltage at any predetermined frequency, without the need for rotating machines. HVDC Light can feed load into a passive network.
- VSC converters are very suitable for creating a DC grid with a large number of converters, since very little coordination is needed between the interconnected VSCs.
- Transient over voltages can be counteracted by fast reactive power response. Thus, stability margins are enhanced because of reactive power support. Moreover, the grid can be operated closer to upper limit, and higher voltage level allows more power transfer.
- Unlike conventional HVDC converters, VSC can operate at very low power. As the active and reactive powers are controlled independently, even at zero active power, the full range of reactive power can be utilized. Active power transfer can be quickly reversed without any change of control mode and without any filter switching or converter blocking. The power reversal is obtained by changing the direction of the DC current and not by changing the DC voltage as for conventional HVDC.

2.5 Applications of Classic HVDC Systems

The first application for classic HVDC systems was to provide point-to-point electrical power interconnections between asynchronous AC power networks. There are other applications that can be met by HVDC converter transmission, which include:

- Interconnections between asynchronous systems. Some continental electric power systems consisting of asynchronous networks such as two grid different frequency (50 HZ and 60 HZ).
- Deliver energy from remote energy sources. Where generation has been developed at remote sites of available energy, HVDC transmission has been an economical means to bring the electricity to load centers. The main application has been the connection of remote hydro-stations to load centers.
- Import electric energy into congested load areas. In areas where new generation is impossible, to bring into service to meet load growth or replace inefficient or decommissioned plant, underground DC cable transmission is a viable means to import electricity.
- Increasing the capacity of existing AC transmission by conversion to DC transmission. New transmission rights of way may be impossible to obtain. Existing overhead AC transmission lines upgraded to or overbuilt with DC transmission can substantially increase the power transfer capability on the existing right of way.
- Power flow control. AC networks do not easily accommodate desired power flow control. Power marketers and system operators may require the power flow control capability provided by HVDC transmission.
- Stabilization of electric power networks. Some widespread AC power system networks operate at stability limits well below the thermal capacity of their transmission conductors. HVDC transmission is an option to consider increasing utilization of network conductors along with the various power electronic controllers, which can be applied on AC transmission.
- An HVDC transmission line has lower losses than AC lines for the same power capacity. The losses in the converter stations have of course to be added, but above a certain break-even distance, the total HVDC transmission losses become lower than the AC losses. HVDC cables also have lower losses than AC cables.

Although thyristor-based HVDC systems represent mature technology, there are still exciting developments worth mentioning such as:

- Active AC and DC filtering.
- Capacitor commutated converter (CCC)-based systems [32].
- An improvement in the thyristor-based commutation, the CCC concept is characterized by the use of commutation capacitors inserted in series between the converter transformers and the thyristor valves. The commutation capacitors reduce the risk of commutation failure of the converters when connected to weak networks.
- Air-insulated outdoor thyristor valves.
- New and advanced cabling technology.

Direct connection of generators to HVDC converters.

Chapter 3
VSC-HVDC Control System

3.1 Introduction

In the case of VSC-based HVDC transmission systems, the transfer of power is controlled in the same way as in the case of a classic HVDC transmission. The inverter side controls the active power, while the rectifier side controls the DC voltage [33, 56]. With classic HVDC, the reactive power cannot be controlled independently of the active power. VSC-HVDC makes it possible to control the reactive power and the active power independently. The reactive power flow can be controlled separately in each converter by the AC voltage that is requested or set manually without changing the DC voltage. The active power flow can be controlled by DC voltage on the DC side or the variation of frequency of AC side or set manually. Thus, the active power flow, the reactive power flow, the AC voltage, the DC voltage, and the frequency can be controlled when using VSC-HVDC. Figure 3.1 represents the control structure of the VSC-HVDC transmission system.

3.2 Control Strategies

Different control strategies are found in literature for the control of VSC-HVDC. One of the methods for control of VSC-HVDC is known as direct power control (DPC) method. DPC is based on the instantaneous active and reactive power control loops [34]. In DPC, there are no internal current control loops and no PWM modulator block, because the converter switching states are selected by a switching table based on the instantaneous errors between the commanded and estimated values of active and reactive power. This method uses estimated virtual flux vector in its control loop. Due to disadvantages like variable switching frequency and necessity of fast conversion and computation, the use of this type of control is not

N. F. Ibrahim, S. S. Dessouky, *Design and Implementation of Voltage Source Converters in HVDC Systems*, Power Systems,
https://doi.org/10.1007/978-3-030-51661-1_3

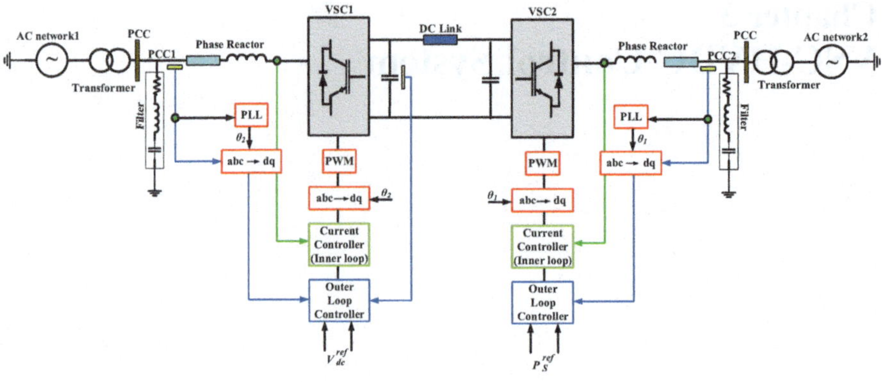

Fig. 3.1 Control structure of the VSC-HVDC transmission system

very common. In addition, call power angle control, which is also called voltage angle control, is perhaps the most straightforward controller for grid-connected VSCs [35].

Another widely used method for control of VSC-HVDC is the vector control method. Vector control method uses modeling of three-phase systems by using axis transformations. Vector theory is most widely used in the control of three-phase PWM converters these days. In PWM converters for AC applications, vector control systems can be utilized to obtain independent control of the active and reactive powers. One of the most advantageous characteristics of vector control is that vectors of AC currents and voltages occur as constant vectors in steady state, and hence, static errors in the control system can be avoided by using PI controllers and finally is hysteresis current controller (HCC) this method is used for pulse generation in current-controlled VSCs.

3.2.1 The Principle of Power Angle Control

The principle of power angle control is based on equations:

$$P = \frac{V_1 V_2 \sin\theta}{X} \tag{3.1}$$

$$Q = \frac{{V_1}^2 - V_1 V_2 \cos\theta}{X} \tag{3.2}$$

where P and Q are the active and reactive power in the AC system with voltage magnitudes V_1 and V_2 between two electrical nodes. The variables θ and X are the phase angle difference and line reactance between the two nodes, respectively [36]. Following equation (3.1), it is straightforward that the active power is mainly associated with the phase angle θ, while the reactive power is more associated with the voltage magnitude difference. The provided mathematical relationships are the base of power angle control. Thus, the active power is controlled by adjusting the voltage phase angle of the VSC. Similarly, the reactive power is controlled by varying the voltage magnitude of the VSC. Despite the fact that power angle control seems to be simple and beneficial, it has never been used in any existing VSC-HVDC systems since it has several severe drawbacks:

- The controller bandwidth is seriously limited.
- The control system is not able to limit the valve current of the converter.

The latter is a severe problem, since the VSC-HVDC converters usually do not provide an overcurrent protection feature. Thus, it is vital for the control system to be able to limit the valve current in order to prevent the converters from being tripped during perturbations [37].

3.2.2 Vector Control Principle

Vector control system involves simplified representation of three-phase systems known as dq transformations.

3.2.3 dq Transformation

The dq is transformation of coordinates from the three-phase stationary coordinate system to the dq rotating coordinate system. This transformation is made in two steps:

- A transformation from the three-phase stationary coordinate system to the two-phase, $\alpha\beta$ stationary coordinate system
- A transformation from the $\alpha\beta$ stationary coordinate system to the dq rotating coordinate system

Clarke and inverse Clarke transformations are used to convert the variables (e.g., phase values of voltages and currents) into stationary $\alpha\beta$ reference frame, and vice versa. Similarly, Park and inverse Park transformations convert the values from stationary $\alpha\beta$ reference frame to synchronously rotating dq reference frame, and vice versa. The reference frames and transformations are shown in Fig. 3.2.

The stationary α-axis is chosen to be aligned with stationary three-phase a-axis for simplified analysis. The dq reference frame is rotating at synchronous speed ω

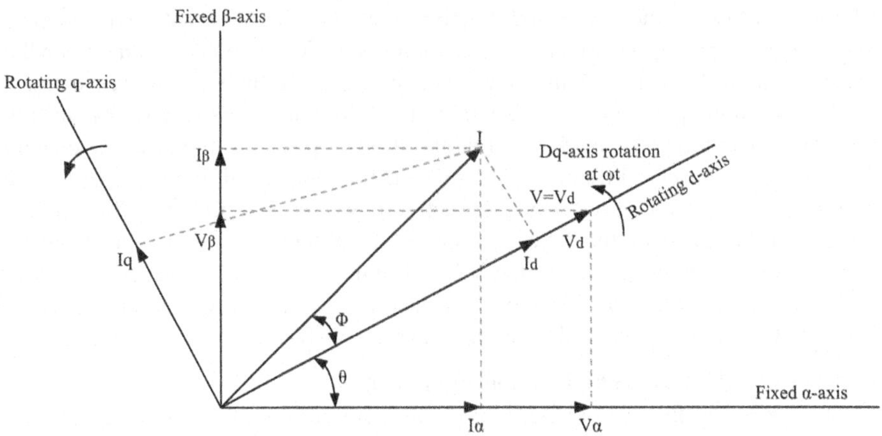

Fig. 3.2 Transformation of axes for vector control

with respect to the stationary frame $\alpha\beta$, and at any instant, the position of *d*-axis with respect to α-axis is given by $\theta = \omega t$.

3.2.4 Vector Control

For analysis of the voltage source converter using vector control, three-phase currents and voltages are described as vectors in a complex reference frame, called $\alpha\beta$ frame. A rotating reference frame synchronized with the AC grid is also introduced, as in Fig. 3.2. As the dq frame is synchronized to the grid, the voltages and currents occur as constant vectors in the dq reference frame in steady state. For the analysis of the system, basic equations describing the system behavior are presented based on analysis done in [38, 39]. The vector control scheme is shown in Fig. 3.3.

By utilizing synchronously rotating dq reference frame, independent active and reactive power control is possible. Initially, system currents and voltages are described as vectors in a stationary $\alpha\beta$ reference frame, and then they are transformed to the rotating dq coordinate system. The transformation to dq coordinates is done as follows: three-phase components $x_a(t)$, $x_b(t)$, and $x_c(t)$ are first described as two vectors in the $\alpha\beta$ reference frame (Fig. 3.4) using Clarke transformation:

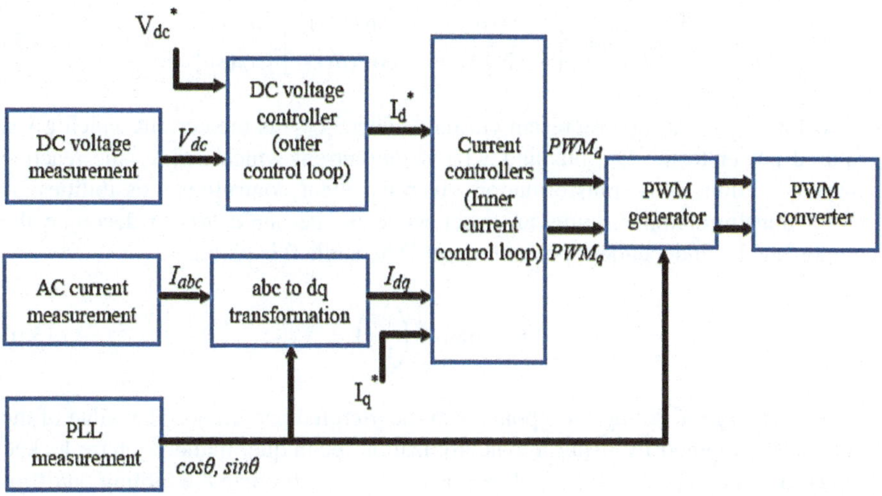

Fig. 3.3 Vector control principle [40]

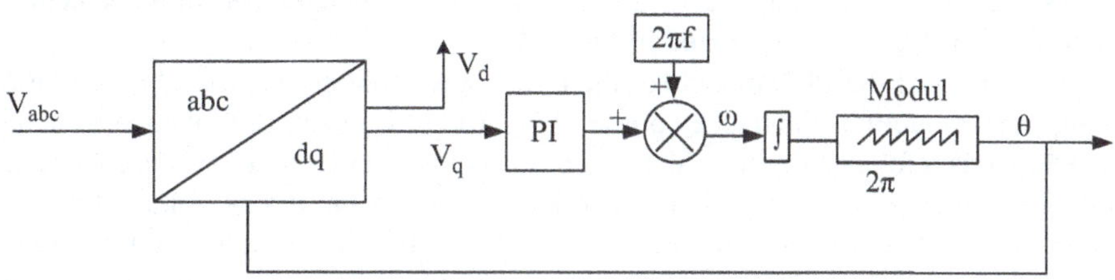

Fig. 3.4 Block diagram of PLL

$$\begin{bmatrix} x_\alpha(t) \\ x_\beta(t) \end{bmatrix} = \begin{bmatrix} \sqrt{\frac{2}{3}} & \frac{-1}{\sqrt{6}} & \frac{-1}{\sqrt{6}} \\ 0 & \frac{1}{\sqrt{2}} & \frac{-1}{\sqrt{2}} \end{bmatrix} \begin{bmatrix} x_a(t) \\ x_b(t) \\ x_c(t) \end{bmatrix} \tag{3.3}$$

where k is a constant coefficient. Using a Park transformation, the transformation from $\alpha\beta$ frame to dq can be written as:

$$x_{dq} = x_{\alpha\beta} e^{-j\theta} \tag{3.4}$$

The vectors $x_\alpha(t)$ and $x_\beta(t)$ are rotating with the angular frequency $\omega(t)$, which is the angular frequency of the grid voltage in rad/s. Let θ(t) be the angle defined by integrating $\omega(t)$ [40]. Then the expanded matrix form of Park transformation is obtained as:

$$\begin{bmatrix} x_{\mathrm{d}}(t) \\ x_{\mathrm{q}}(t) \end{bmatrix} = \begin{bmatrix} \cos(\theta(t)) & \sin(\theta(t)) \\ -\sin(\theta(t)) & \cos(\theta(t)) \end{bmatrix} \begin{bmatrix} x_{\alpha}(t) \\ x_{\beta}(t) \end{bmatrix} \tag{3.5}$$

Vectors $x_{\mathrm{d}}(t)$ and $x_{\mathrm{q}}(t)$ represent currents, where $x_{\mathrm{d}}(t)$ is the current which gives required power to the DC bus and $x_{\mathrm{q}}(t)$ is the current which defines the reactive power condition. This transformation gives excellent controlling possibilities. A correct transformation requires an exact value of the angle $\theta(t)$ to decouple the components for independent power control. The angle θ is given as:

$$\theta = \tan^{-1}\left(\frac{v_{\beta}}{v_{\alpha}}\right) \tag{3.6}$$

where v_{α} and v_{β} are voltage components in the $\alpha\beta$ reference frame. The value of the angle θ is computed by using a synchronization technique, namely, phase-locked loop (PLL) [41]. The phase-locked loop is used to synchronize the turning on/off of the power devices and calculate and control the flow of active/reactive power by transforming the feedback variables to a reference frame suitable for control purposes [42]. The block diagram of the PLL algorithm implemented in the synchronous reference frame is presented in Fig. 3.4.

The inputs of the PLL model are the three-phase voltages measured on the grid side, and the output is the tracked phase angle. The PLL model is implemented in dq synchronous reference frame, which means that a Park transformation is needed. The phase locking of this system is realized by controlling the q-axis voltage to zero. Normally, a PI controller is used for this purpose. By integrating the sum between the PI output and the reference frequency, the phase angle is obtained. The transfer function of the dq PLL system is given by [43]:

$$H(s) = \frac{K_{\mathrm{p.s}} + \frac{K_{\mathrm{p}}}{T_{\mathrm{i}}}}{s^2 + K_{\mathrm{p.s}} + \frac{K_{\mathrm{p}}}{T_{\mathrm{s}}}} \tag{3.7}$$

As it can be observed, this equation is similar to the second order transfer function having a zero that is shown in the following expression:

$$G(s) = \frac{2\zeta\omega_{\mathrm{n.s}} + {\omega_{\mathrm{n}}}^2}{s^2 + 2\zeta\omega_{\mathrm{n.s}} + {\omega_{\mathrm{n}}}^2} \tag{3.8}$$

Comparing the equations (3.7) and (3.8) can obtain the gain of the controller. In order to calculate the parameters of the controller, a settling time T_{set} of 0:04 s and a damping factor $\zeta = \frac{1}{\sqrt{2}}$ are chosen. The PI parameters can be calculated as:

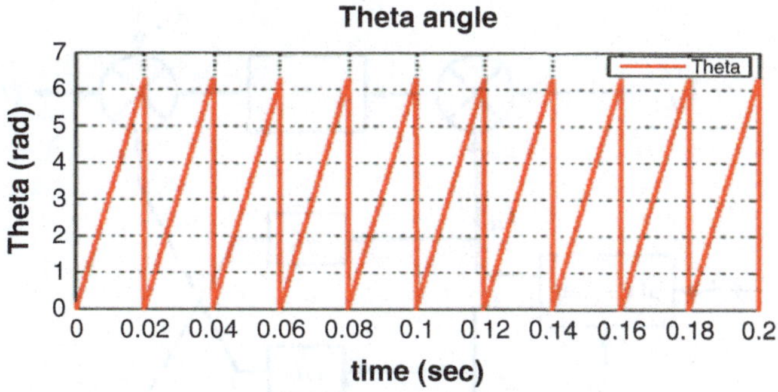

Fig. 3.5 Phase angle of the grid voltage

$$K_p = 2\zeta\omega_n = \frac{9.2}{T_{set}} \tag{3.9}$$

$$T_i = \frac{T_{set}\zeta^2}{2.3} \tag{3.10}$$

where the natural frequency, ω_n, is given by:

$$\omega_n = \frac{4.6}{\zeta T_{set}} \tag{3.11}$$

The grid phase angle obtained with the described PLL algorithm is shown in Fig. 3.5.

3.2.5 Inner Current Controller

3.2.5.1 System Description and Transfer Functions

The inner current controller is implemented in the dq synchronous reference frame [44]. Usually, the dq control structures are associated with PI controllers due to their good behavior when regulating DC variables. However, according to [45], the PI current controllers have no satisfactory tracking performances when they have to regulate coupled systems like the one described in Equations (3.7) and (3.8). Therefore, in order to improve the performances of the PI current controllers in such systems, cross-coupling terms and voltage feed forward is usually used [46]. The structure of the inner current controller implemented in the synchronous reference frame is presented in Fig. 3.6.

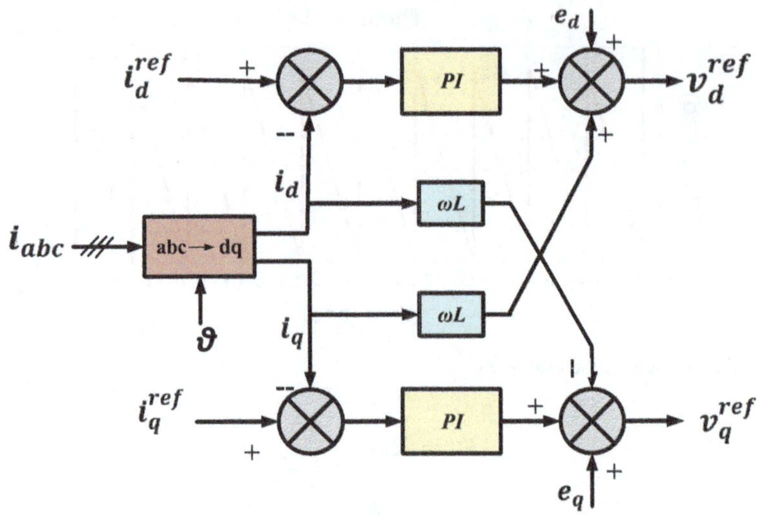

Fig. 3.6 The structure of the inner current controller implemented in synchronous reference frame

3.2.5.2 PI Regulator

The representative equation of the PI regulator is:

$$R(s) = K_p + \frac{K_i}{s} = K_p \cdot \left| \frac{1 + T_{i.s}}{T_{i.s}} \right| \tag{3.12}$$

where the proportional gain K_p and integral time constant $T_i = K_p/K_i$ are the design parameters to be specified. Thus, for PI controller block:

$$\{I_{ref}(s) - I(s)\}\left(K_p + \frac{K_i}{s}\right) = v_{conv}(s) \tag{3.13}$$

3.2.5.3 PWM Converter

From the control point of view, the converter is considered as an ideal power transformer with a time delay. The output voltage of the converter is assumed to follow a voltage reference signal with an average time delay equals half of a switching cycle, due to VSC switches. Hence, the general expression is:

$$Y(s) = \frac{1}{1 + T_{a.s}} \tag{3.14}$$

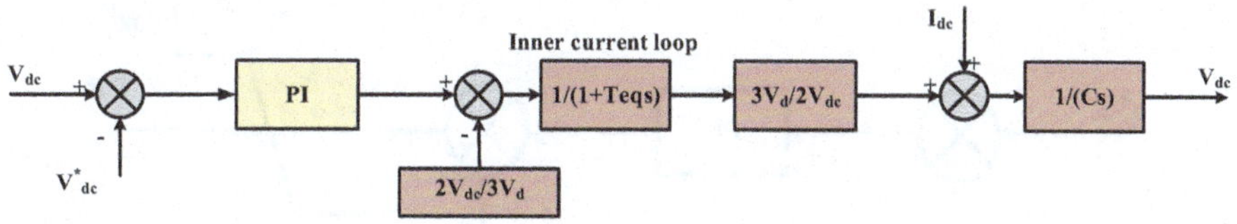

Fig. 3.7 Closed loop control diagram of DC voltage controller [56]

where $T_a = T_{switch}/2$. Thus, for converter block:

$$v'_{conv}(s).\frac{1}{(1+s_{Ta})} = v_{conv}(s) \tag{3.15}$$

3.2.6 *The Outer Controller*

Common objectives for the outer loop active power through the converter, reactive power at each side, and DC-link voltage: Active current (i_d) is used to control active power flow or DC voltage level. Similarly, reactive current (i_q) is used to control reactive power flow into stiff grid connection. The outer controller block diagram is shown in Fig. 3.7.

3.2.7 *DC Voltage Control Using Classic Control Algorithm*

Dimensioning of the DC-link voltage controller is determined by the transfer function between the defined current reference value and the DC-link voltage. From power balance of the VSC-HVDC terminal:

$$P_{ac} + P_{dc} + P_{cap} = 0 \tag{3.16}$$

$$\frac{2}{3}v_d i_d + V_{dc} * I_{dc} + V_{dc} * i_{cap} = 0 \tag{3.17}$$

where I_{dc} and i_{cap} are the DC bus current and the capacitor current, respectively, and 3/2 factor comes from the Park's transformation. From (3.14), the current through the capacitor is:

$$i_{cap} = -\left(\frac{3v_d i_d}{2V_{dc}} + I_{dc}\right) \tag{3.18}$$

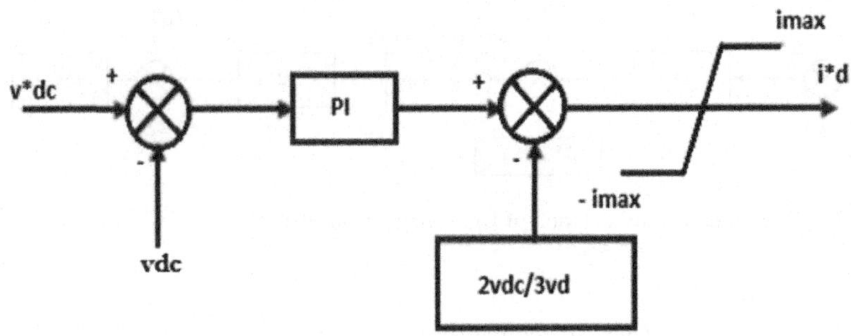

Fig. 3.8 DC voltage regulator block diagram

and the same current in terms of voltage across the capacitor is given by:

$$i_{\mathrm{cap}} = c\frac{dV_{\mathrm{dc}}}{dt} \tag{3.19}$$

From Equations (3.15) and (3.16), the differential equation for the DC voltage becomes

$$\frac{dV_{\mathrm{dc}}}{dt} = -\frac{3v_{\mathrm{d}}i_{\mathrm{d}}}{2c * V_{\mathrm{dc}}}\left(i_{\mathrm{d}} + \frac{2V_{\mathrm{dc}} * I_{\mathrm{dc}}}{3v_{\mathrm{d}}}\right) \tag{3.20}$$

Equation (3.17) shows that DC voltage is regulated by controlling the active current i_{d}. The I_{dc} term from the equation (3.17) is compensated by feed forward in the DC voltage controller. The DC voltage control block diagram is presented below. From Equation (3.16), the block diagram seen in Fig. 3.8 is obtained.

3.2.8 *Active and Reactive Power Controllers*

The value representation of active and reactive power of the system in dq transformation can be written as:

$$P = v_{\mathrm{d}} * i_{\mathrm{d}} \tag{3.21}$$

$$q = -v_{\mathrm{q}} * i_{\mathrm{q}} \tag{3.22}$$

More accurate active and reactive power control is achieved if using PI controllers a feedback loop is employed. Active and reactive power control loops are shown in Fig. 3.9.

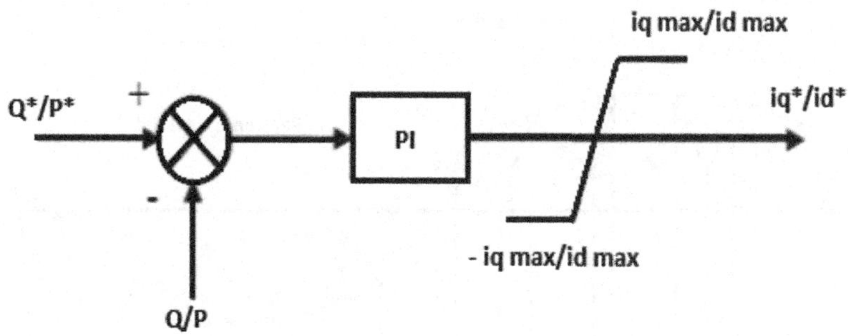

Fig. 3.9 Block diagram of active power control

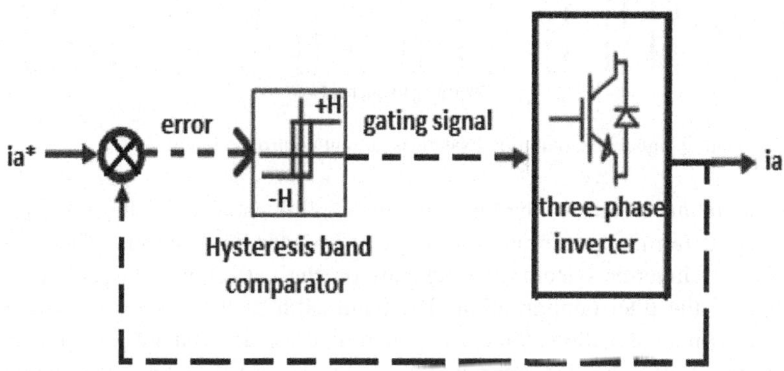

Fig. 3.10 Block diagram of current hysteresis control technique

3.3 Hysteresis Current Control

The hysteresis current control (HCC) technique is used for pulse generation in current-controlled VSCs. The control method offers good stability, gives a very fast response, provides good accuracy, and has got a simple operation [47]. The control of inverter can be realized by the hysteresis control technique. It is composed of a hysteresis around the reference line current. The reference line current of the inverter is referred to as (i_a), and the actual line current of the inverter is referred to as (i_a*). This control scheme is shown in Fig. 3.10. In this control scheme, a signal deviation (H) is designed and imposed on i_a* to form the upper and lower limits of a hysteresis band. Hysteresis current controller with a fixed H. The switching frequency of the hysteresis band current control method described above depends on how fast the current changes from the upper limit of the hysteresis band to the lower limit of the hysteresis band, or vice versa.

The rate of change of the actual inverter line currents varies the switching frequency; therefore, the switching frequency does not remain constant throughout the switching operation, but varies along with the current waveform. Furthermore,

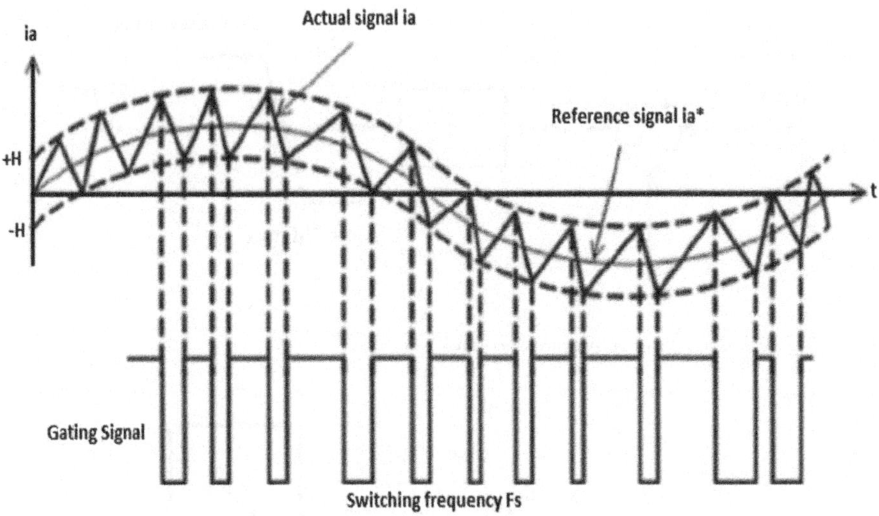

Fig. 3.11 Gating signal generation by hysteresis current control technique [48]

the line inductance value of the filter and the DC link capacitor voltage are the main parameters determining the rate of change of inverter line currents. The switching frequency of inverter system also depends on the capacitor voltage and the line inductors of the filter configuration. The bandwidth of the hysteresis current controller determines the allowable current shaping error. By changing the bandwidth, the user can control the average switching frequency of the inverter and evaluate the performance for different values of hysteresis bandwidth. In principle, increasing the inverter operating frequency helps to get a better compensating current waveform [48]. However, there are device limitations, and increasing the switching frequency causes increasing losses on the switching transistors.

The advantages of using the current hysteresis control are its excellent dynamic performance and controllability of the peak-to-peak current ripple within a specified hysteresis. Figure 3.11 shows the gating signal generated by current hysteresis control technique. An error signal (e) is used to control the switches in an inverter. This error is the difference between the desired inverter current (i_a) and its estimated reference current signal (i_a*) as $e = \Delta I = i_a - i_a^*$.

When the error reaches an upper limit as $\Delta I \geq H/2$, the transistors are switched to force the current down. When the error reaches a lower limit as $\Delta I \leq -H/2$, the current is forced to increase.

3.4 Filter Design

Recently, the development of renewable energy technologies has been accelerating, making the simultaneous development of power conversion devices for applications, such as wind and PV systems, extremely important, and the development of these technologies is actively underway. The harmonics caused by the switching of the power conversion devices are the main factor-causing problems to sensitive equipment or the connected loads, especially for applications above several kilowatts, where the price of filters and total harmonic distortion (THD) is also an important consideration in the systems design phase [49]. The inductance of the input or output circuits of the power conversion devices has conventionally been used to reduce these harmonics. However, as the capacity of the systems has been increasing, high values of inductances are needed, so that realizing practical filters has been becoming even more difficult due to the price rises and the poor dynamic responses.

An L filter or LCL filter is usually placed between the inverter and the grid to attenuate the switching frequency harmonics produced by the grid-connected inverter as shown in Fig. 3.12.

Compared with L filter, LCL filter has better attenuation capacity of high-order harmonics and better dynamic characteristic [50]. However, an LCL filter can cause stability problems due to the undesired resonance caused by zero impedance at certain frequencies. To avoid this resonance from contaminating the system, several damping techniques have been proposed. One way is to incorporate a physical passive element, such as a resistor in series with the filter capacitor [51] as shown in Fig. 3.13. This passive damping solution is very simple and highly reliable.

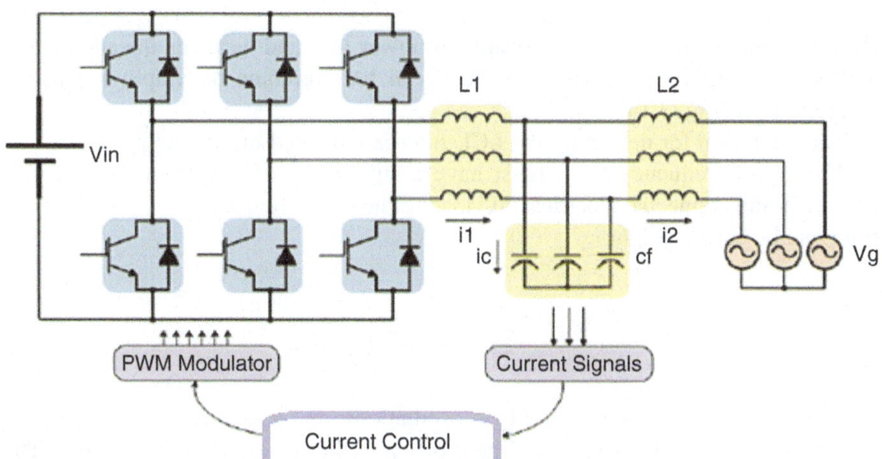

Fig. 3.12 LCL filter-based grid-connected three-phase inverter [49]

Fig. 3.13 LCL filter and components

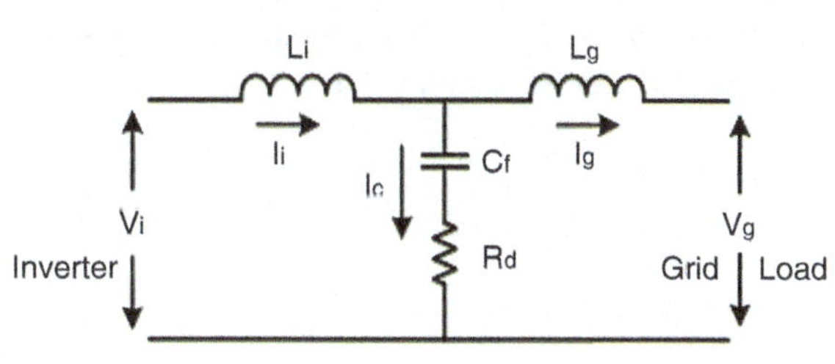

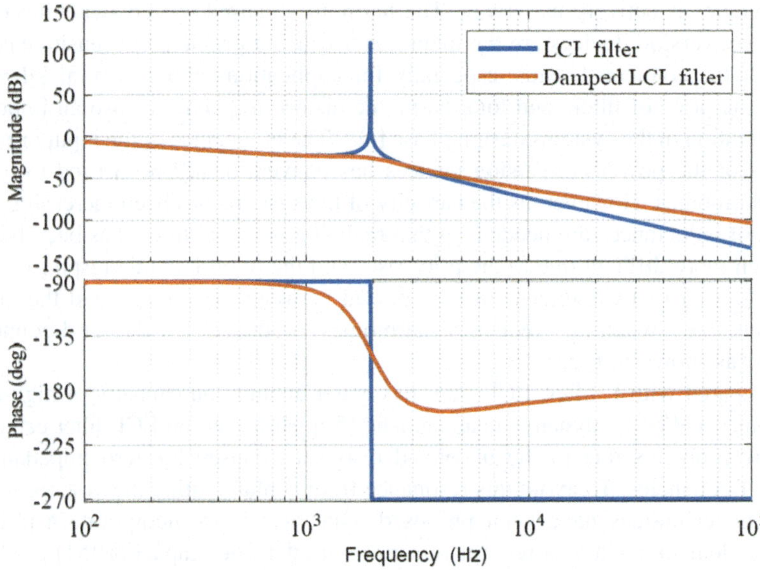

Fig. 3.14 LCL filter with passive damping resistance

However, the additional resistor results in power loss and weakens the attenuation ability of the LCL filter. This drawback can be overcome by employing active damping [32] (Fig. 3.14).

The algorithm for designing the LCL filter is indicated in Fig. 3.15.

The cutoff frequency (f_{res}) must have a sufficient distance from the grid frequency or the connected electrical AC loads. The cutoff frequency (f_{res}) of the LCL filter can be calculated as:

$$f_{\text{res}} = \frac{1}{2\pi} \times \sqrt{\frac{L_{\text{i}} + L_{\text{g}}}{L_{\text{i}} \times L_{\text{g}} \times C_{\text{f}}}} \tag{3.23}$$

The LCL filter will be vulnerable to oscillations, and it will magnify frequencies around its cutoff frequency (f_{res}). Therefore, the filter is added with damping. The simplest way is to add damping resistor (R_{d}). The variant with resistor connected in series with the filter capacitor has been chosen [33]. The passive damped LCL filter frequency response is shown in Fig. 3.15. However, it is obvious that the damping resistor reduces the efficiency of the overall system. The value of the damping resistor (R_{d}) can be calculated as:

$$R_{\text{d}} = \frac{1}{3\omega_{\text{res}} C_{\text{f}}} \tag{3.24}$$

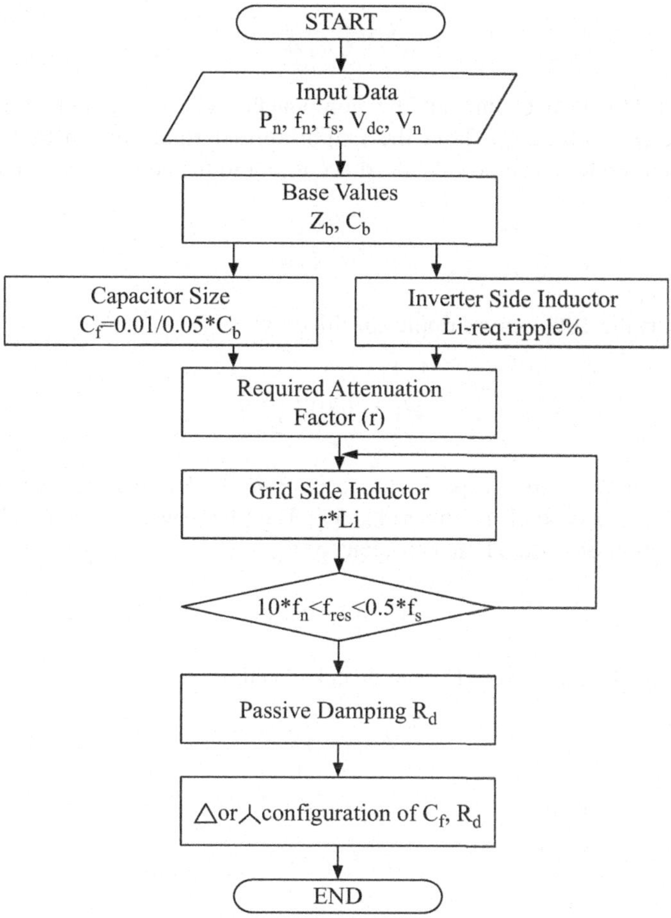

Fig. 3.15 LCL filter design algorithm

The following parameters are needed for the filter design:

U_n – line-to-line RMS voltage (inverter output)
P_n – rated active power
V_{dc} – DC link voltage
f_n – grid frequency
f_s – switching frequency

Thus, the filter values will be referred to in a percentage of the base values [49, 50]:

$$Z_b = U_n^2 / P_n \tag{3.25}$$

$$C_b = \frac{1}{\omega_n Z_b} \tag{3.26}$$

The first step in calculating the filter components is the design of the inverter side inductance (L_i), which can limit the output current ripple by up to 10% of the nominal amplitude. It can be calculated according to the equation derived in [51]:

$$L_i = \frac{V_{dc}}{16\, f_s\, \Delta I_L} \tag{3.27}$$

where ΔI_L is the 10% current ripple specified by:

$$\Delta I_L = 0.01 \frac{P_n \sqrt{2}}{U_n} \tag{3.28}$$

The design of the filter capacity proceeds from the fact that the maximal power factor variation acceptable by the grid is 5%. The filter capacity can be calculated as a multiplication of system base capacitance (C_b):

$$C_f = 0.05\, C_b \tag{3.29}$$

The grid-side inductance (L_g) can be calculated as:

$$L_g = r * L_i \tag{3.30}$$

where (r) is the factor between (L_i) and (L_g).

Chapter 4
VSC-HVDC Under AC and DC Fault Conditions

4.1 Introduction

This chapter will examine high-voltage direct current (HVDC) system responses to DC and AC faults from a theoretical perspective. The analysis will consider the HVDC topology two-level voltage source converter (VSC). The test system is shown in Fig. 4.1, where the location of faults is also illustrated. The investigation is concerned with system responses for faults located at the converter DC or AC bus, which is the worst-case fault location, and it is representative for faults further away from the converter. It is presumed that faults will not happen closer to the converter, because this area is located in the valve hall and faults are unlikely. If faults happen closer to the converter, then the fault current will be too fast and strong for the normal protection methods, and insulated gate bipolar transistors (IGBTs) will be tripped by the internal switch-level protection.

4.2 Faults on the DC System

Cable faults are more common than faults in other parts of the system [35]. This is due to the variety of conditions that can cause a DC fault. The most frequent reason is insulation deterioration or breakdown. However, there are several other reasons that can lead to the same result, such as electrical stresses, environmental conditions, aging, and physical damage. The DC faults that are possible in a HVDC system can be categorized as [49]:

- Positive pole-to-ground fault
- Negative pole-to-ground fault
- Pole-to-pole fault

N. F. Ibrahim, S. S. Dessouky, *Design and Implementation of Voltage Source Converters in HVDC Systems*, Power Systems,
https://doi.org/10.1007/978-3-030-51661-1_4

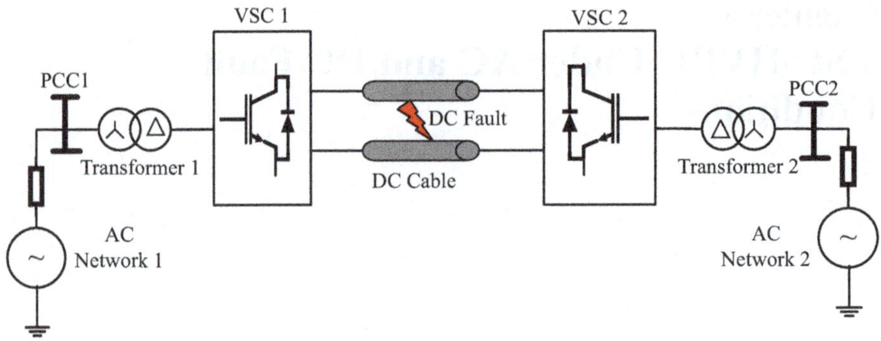

Fig. 4.1 A VSC-HVDC transmission system with a location for DC fault

Since the system is in general symmetrical, negative pole-to-ground fault can be considered a mirror of the positive pole-to-ground fault. Therefore, in the remaining of this book, only the pole-to-pole fault and positive pole-to-ground fault are further investigated. Faults on DC lines are, usually, pole-to-ground faults [35–49]. A pole-to-pole fault happens when the two active conductors are connected either directly or through the ground. This occurs seldom, as it requires a severe damage of both HVDC line cables. Especially regarding submarine cables, where most of the cable problems are created by mechanical stresses, a short-circuit fault is very unlikely.

4.2.1 Line-to-Ground Fault

This type of fault occurs when one line (positive or negative pole) is short-circuited to the ground. This can happen due to the reasons already mentioned. Most commonly, broken insulation, due to environmental and mechanical fatigue, can allow a current path to ground. As the fault persists, it deteriorates the cable insulation and worsens the fault situation. In case of submarine cables, ground faults can happen when ships anchor and cut one of the lines. In case of overhead lines, lightings or construction works can result in a cut line falling to the ground [49, 55]. In all of these cases, the fault is permanent, and the line needs to be completely isolated for the cable to be replaced.

4.2.2 Pole-to-Pole Fault

A pole-to-pole fault is less likely to occur than a line-to-ground fault. It is generally very unlikely, due to the distance between the lines, that an object falls and cuts both the positive and the negative line at the same time, resulting in a short circuit through the ground. Especially in overhead lines, a lightning cannot cause a bipolar fault

[49]. As far as undersea cables are considered, they are almost immune to such faults, as they are most often separated from each other, and are well protected by several layers of insulation and conduit, which make less likely the occurrence of a simultaneous fault that would bring both conductors in touch. The only case that a pole-to-pole fault is more probable is the occurrence of a switching fault or failure at the converter station, which could result in a short circuit between the two poles.

4.3 DC Fault Analysis

In this section, a more elaborate analysis is provided for both types of investigated faults. The theoretically expected VSC system behavior is presented for a better understanding of the complete system reaction. The DC fault analysis is presented in the simulation results of Chap. 5. For the analysis, it is assumed, based on circuit theory, that the DC cable fault is the same as a DC bus fault. In order to better understand the nonlinear nature of the fault, it is divided into stages for each of which the respective equations are provided.

4.3.1 Pole-to-Ground Fault

The pole-to-ground fault analysis depends on the grounding of the DC system and the converter. This determines the loop that will be created between the fault ground and the rest of the system. The converter is usually grounded at the neutral point in a bipolar configuration. Otherwise, in case of monopolar HVDC links, the return can be either through the ground or via a grounded metallic return. In case of a symmetric monopole, the only loop there can be is with the middle point of the DC link. This is generally employed to reduce the imbalance between the positive and negative voltages and currents. The loop formed between the grounding of the DC side and the AC side of the converter can be avoided if an AC transformer is used after the AC filters with no grounding on its secondary winding, which provides galvanic isolation from the AC grid. In the following analysis, a transformer with a ground on its secondary is assumed. Let us consider a positive pole-to-ground fault. The only change in case of a negative pole-to-ground fault will be the direction of the current and the active diodes that feed the fault [43, 54]. Figure 4.2 depicts a positive pole-to-ground fault and only shows the VSC, which is closer to the fault.

4.3.1.1 First Stage

There is a loop formed between the fault ground and the middle point of the DC link. The circuit can be analyzed with equivalent circuits that show the different stages of fault development. It should also be noted that the fault resistance value plays a role

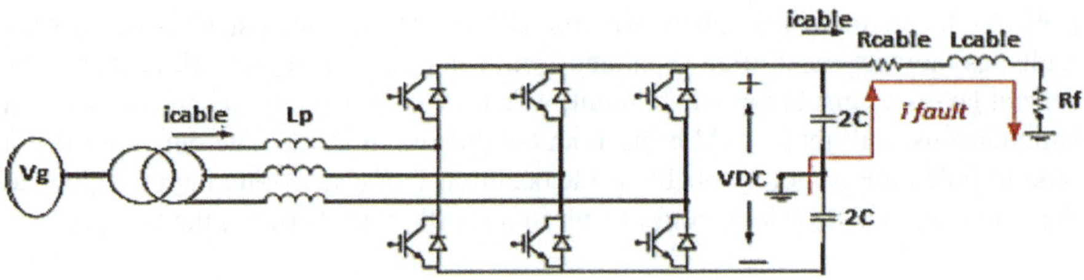

Fig. 4.2 VSC with positive pole-to-ground fault

Fig. 4.3 DC capacitor discharge stage

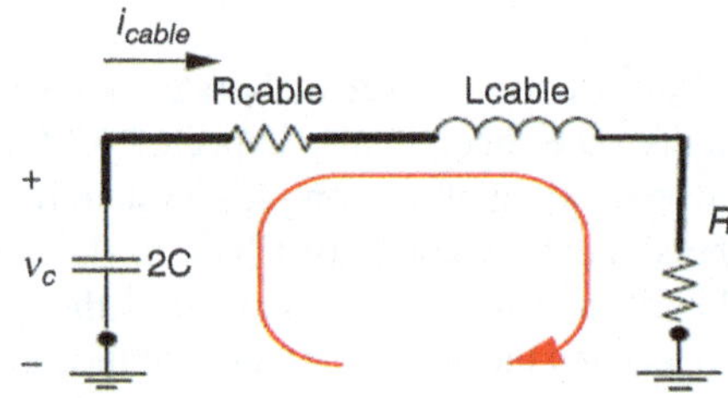

in the oscillation damping. First, the positive pole capacitor gets discharged through the fault, contributing significantly to the fault peak current. This stage is a series RLC circuit with certain initial conditions, $V_C(t_0) = V_0$ and $i_{\text{cable}}\ (t_0) = I_0$. It can be assumed that the discharge takes place as a natural response, without the current contribution from the converter. Figure 4.3 presents the equivalent circuit.

$$\frac{d^2V_C}{dt^2} + 2a\frac{dV_c}{dt} + \omega^2{}_0V_c = 0 \tag{4.1}$$

where $a = \frac{R}{2L}$ and $\omega_0 = \sqrt{\frac{1}{LC}}$. In case of a ground fault, the value of resistance R for the circuit becomes $R = R_{\text{cable}} + R_{\text{f}}$, and thus, $a = \frac{R_{\text{cable}}+R_{\text{f}}}{2l_{\text{cable}}}$ and $\omega_0 = \sqrt{\frac{1}{l_{\text{cable}}2c}}$. The solution of this equation depends on the relation between α and ω_0. Therefore, for a complete analysis of pole-to-ground fault.

4.3.1.2 Second Stage

The second stage begins as soon as the voltage of the capacitor falls below any grid voltage. Either the IGBTs lose control due to overcurrent or they are blocked for the sake of protection. The capacitor voltage collapses, and it is bypassed by the freewheeling diodes of the converter. This is a forced response, which can create a significant overcurrent that can damage the diodes and subsequently the VSC [49, 56]. VSC-HVDC stations have very weak overload capability in contrast to classical HVDC, and it is generally admitted that a VSC converter can handle up to 2 pu current for a limited time, i.e., only a few milliseconds. Figure 4.4 shows the

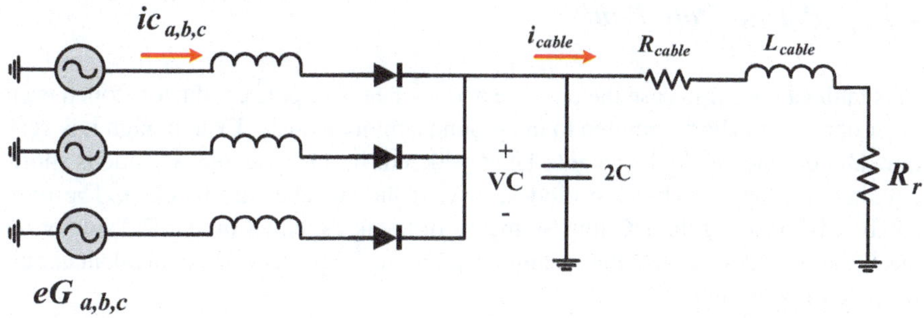

Fig. 4.4 Grid current feeding stage

grid current feeding stage. This stage does not have a linear response, while a continuous solution demands knowledge of the state variables of the previous diode conduction period as initial conditions. The system of state space equations that can best describe this phase is [35]:

$$\begin{vmatrix} \frac{dV_c}{dt} \\ \frac{dI_{cable}}{dt} \\ \frac{di_c}{dt} \end{vmatrix} = \begin{vmatrix} 0 & -\frac{1}{2c} & \frac{1}{2c} \\ \frac{1}{L_{cable}} & -\frac{R_{cable}+R_f}{L_{cable}} & 0 \\ -\frac{1}{Lp} & 0 & 0 \end{vmatrix} \begin{vmatrix} V_c \\ I_{cable} \\ i_c \end{vmatrix} + \begin{vmatrix} 0 \\ 0 \\ 1/Lp \end{vmatrix} eG_{a,b,c} \tag{4.2}$$

4.3.1.3 Third Stage Finally

The third stage includes the steady-state response in case no protection is used. The grid continues feeding the fault infinitely with equivalent impedance [35]:

$$Z = (R_f + R_{cable} + j\omega s L_{cable}) \| \left(\frac{1}{j\omega s 2c}\right) + j\omega s L_P = |Z| \lfloor k \tag{4.3}$$

where ωs is the synchronous angular frequency and L_P is the grid-side inductance. The current through each diode is then given by:

$$i_{D1} = i_{ca}, (>0) = \frac{e_G \lfloor \theta}{|Z| \lfloor k} = \frac{e_G}{|Z|} \lfloor \theta - k \tag{4.4}$$

where θ is the phase angle of the grid voltage. The cable current is subsequently given by:

$$i_{cable} = iD_1 + iD_2 + iD_3 \tag{4.5}$$

where D_1, D_2, and D_3 are the positive phase arm diodes.

4.3.2 Pole-to-Pole Fault

This fault can occur in case the positive and the negative pole conductors touch each other or are both short-circuited to the ground simultaneously. Even though it is very rare, all possible DC faults have to be considered. In this case, the AC side is short-circuited via the freewheeling diodes, even if the switches are blocked. The only solution is to isolate the DC line by means of breakers, either on the DC side or on the AC side, to de-energize the system [49]. Figure 4.5 presents the equivalent circuit for pole-to-pole fault.

4.3.2.1 First Stage

The first stage of the fault is the same as for the previously analyzed DC fault type. The DC capacitor is naturally discharged through an RLC circuit [43]. Based on the line parameters, $\alpha < \omega_0$, and thus, the second-order circuit has the same under-damped response as in the case of pole-to-ground fault. Equations (4.1) and (4.2) apply in this case with resistance R only taking into account the line resistance; no fault resistance is present in short circuit. The fault current is given by (4.6):

$$i_{\text{cable}} = I_0 e^{-(R_{\text{cable}}/L_{\text{cable}})t} \tag{4.6}$$

where R_{cable} and L_{cable} are the line resistance and inductance from the converter to the fault point, respectively, and I_0 is the initial value of the cable current at the fault moment. The fault current increases to a high peak and then decreases exponentially to zero, and also the DC-link voltage reduces rapidly to zero.

4.3.2.2 Second Stage

During the second stage, the DC-link voltage has dropped to zero, and the cable inductance is discharged through the freewheel path as shown in Fig. 4.6.

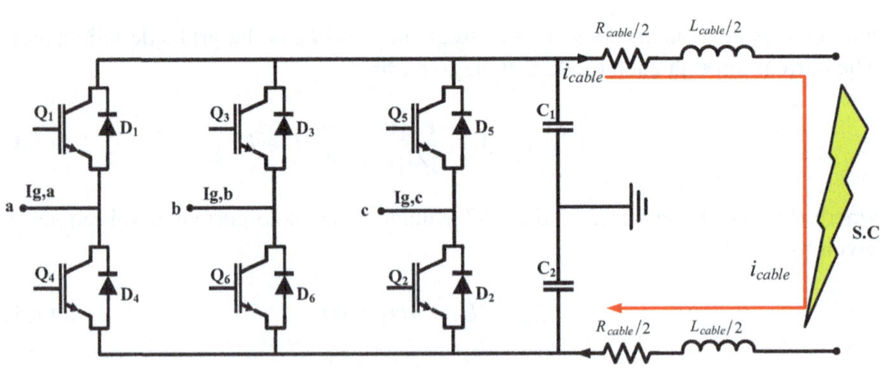

Fig. 4.5 DC pole-to-pole fault equivalent circuit

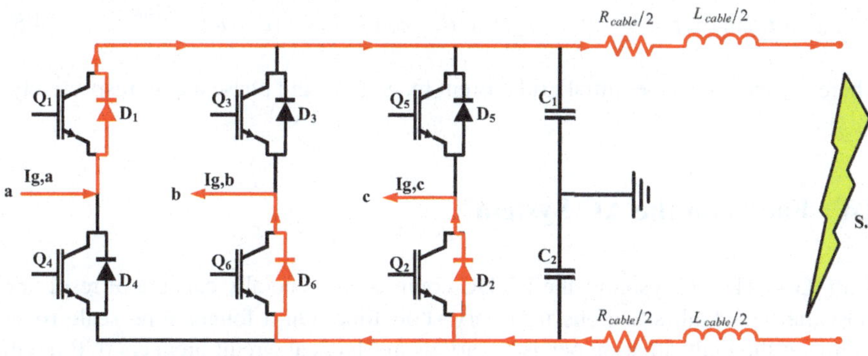

Fig. 4.6 Diode freewheel equivalent circuit

Fig. 4.7 Equivalent circuit of capacitor recharging stage

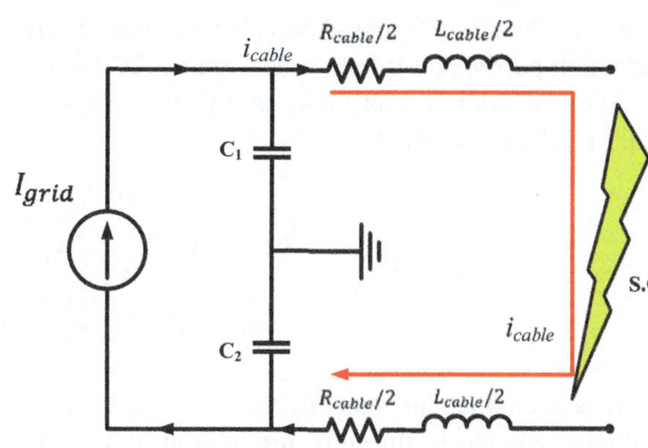

4.3.2.3 Third Stage Finally

The last stage includes a forced current source response. Figure 4.7 depicts the system during this phase. The VSC station functions as a current source, which feeds the fault. Finally, the system reaches a steady state [43, 56]. The grid phase voltage and current after the fault is:

$$\begin{aligned} e_g &= E_g \sin(\omega_s t + \theta) \\ i_g &= I_g \sin(\omega_s t + \theta - \varphi) + I_0 e^{-\left(\frac{R_{\text{cable}}}{L_{\text{Cable}}}\right)t} \end{aligned} \tag{4.7}$$

where E_g is the phase voltage magnitude, I_g is the grid current magnitude, Θ is the grid angle, ϕ is the phase angle, and ω_s the angular frequency. Considering the initial conditions at the fault moment, Eq. (4.8) can be written as:

$$i_g = I_g \sin(\omega_s t + \theta - \varphi) + I_{g0} \sin(\theta - \varphi_0) - I_0 \sin(\theta - \varphi) e^{-\left(\frac{R_{cable}}{L_{cable}}\right)t} \quad (4.8)$$

where I_{g0} and ϕ_0 are the initial grid current magnitude and phase angle, respectively.

4.4 Faults on the AC System

With IGBT HVDC systems, the HVDC controls are generally capable of resolving most transient fault situations in a very short time. On a longer time scale (over 50 ms), if the fault situation persists, various mechanical circuit breakers (CBs) will react. A number of mechanical CBs are associated with an HVDC system, but they are used only as a final means of protection, that is, when HVDC controls can solve disturbance and permanent isolation is required. The tripping of CBs implies loss of capacity. Figure 4.8 shows some typical fault locations. The HVDC controls always react to a disturbance first. A tap changer may also react when a modest AC voltage deviation persists for a long time (remote AC fault or tripping).

The converter current under an extreme AC system fault can be obtained from Eq. (4.9) by replacing $V_g = 0$:

$$Ig_d = \frac{-V_{cq}}{X_t} = \frac{-M_q V_{dc}}{2X_t} \quad (4.9)$$

$$Ig_q = \frac{V_{cd}}{X_t} = \frac{M_d V_{dc}}{2X_t} \quad (4.10)$$

It is seen that, in general, the converter can control an AC fault current by reducing voltage V_c through current control loops (using M_d and M_q). This current reduction is achieved by reducing significantly the modulation index M_d, which is typically much larger than M_q. The DC voltage may remain at rated value during the AC fault. However, the converter control loop has limited speed as it is based on feedback measurements and involves signal processing. In the first instant after the

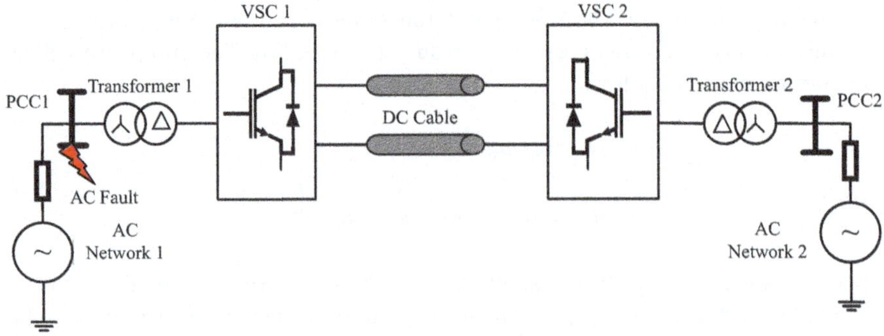

Fig. 4.8 Fault locations to HVDC systems

fault, assuming that converter DC voltage is at rated value, only the reactor/transformer (X_t) limits the current. A VSC-HVDC is normally designed to ride through all AC faults without tripping. If an AC fault persists for a long time, the converter is tripped. The current peak depends primarily on the speed of inner control loops, but it is also significantly influenced by the transformer reactance, and this is one of the main requirements in selecting X_t.

4.4.1 *Transient Grid Faults: the Voltage Sags (Dips)*

A voltage sag, also called a voltage dip, is a sudden reduction of the grid voltage at the PCC, generally between 10% and 90% of the rated value, during a period lasting from half a cycle to a few seconds. Voltage sags usually happen as a consequence of short circuits, faults to ground, transformer energizing, and connection of large induction motors. Depending on both the type of grid fault and the transformer connections along the power lines, it is possible to distinguish between different types of voltage sags. The definition of guidelines to classify voltage sags is a matter that remains under discussion still today [44, 53]. The $\vec{D}\overrightarrow{Vsa^+}$ is known as the characteristic voltage of the voltage sag and represents either the phase voltage in phase-to-ground faults or the line-to-line voltage in phase-to-phase faults. Likewise, the phase angle of $\vec{D}$ is known as the "characteristic phase angle jump" of the voltage sag. Figures 4.9, 4.10, 4.11, and 4.12 show four types of voltage sag.

4.4.1.1 Sag Type A: Three-Phase Fault and Three-Phase-to-Ground Fault

$$\begin{bmatrix} \overrightarrow{va} \\ \overrightarrow{vb} \\ \overrightarrow{vc} \end{bmatrix} = \begin{bmatrix} \cdot & 1 & \cdot \\ -\frac{1}{2} & \cdot & -j\frac{\sqrt{3}}{2} \\ -\frac{1}{2} & \cdot & +j\frac{\sqrt{3}}{2} \end{bmatrix} \vec{D}\,\overrightarrow{Vsa^+} \tag{4.11}$$

$$\begin{bmatrix} \overrightarrow{va+} \\ \overrightarrow{va-} \\ \overrightarrow{va0} \end{bmatrix} = \begin{bmatrix} 1 \\ 0 \\ 0 \end{bmatrix} \vec{D}\,\overrightarrow{Vsa^+} \tag{4.12}$$

$$\vec{D} = \frac{\overrightarrow{Zf+}}{\overrightarrow{Zs} + \overrightarrow{Zf+}} \tag{4.13}$$

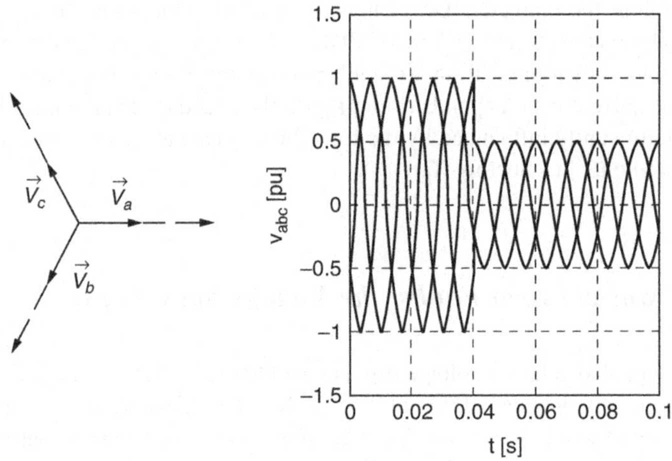

Fig. 4.9 Sag type A [44]

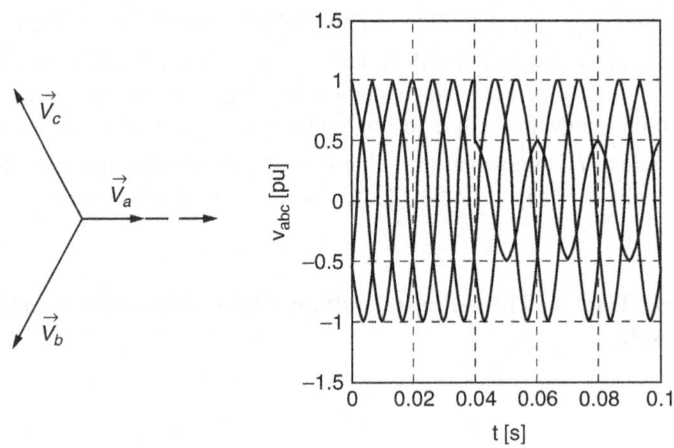

Fig. 4.10 Sag type B [44]

4.4.1.2 Sag Type B: Single-Phase-to-Ground Fault

$$\begin{bmatrix} \overrightarrow{\mathrm{va}} \\ \overrightarrow{\mathrm{vb}} \\ \overrightarrow{\mathrm{vc}} \end{bmatrix} = \begin{bmatrix} \cdot & \vec{D} & \cdot \\ -\frac{1}{2} & \cdot & -\mathrm{j}\frac{\sqrt{3}}{2} \\ -\frac{1}{2} & \cdot & +\mathrm{j}\frac{\sqrt{3}}{2} \end{bmatrix} \overrightarrow{Vsa^{+}} \tag{4.14}$$

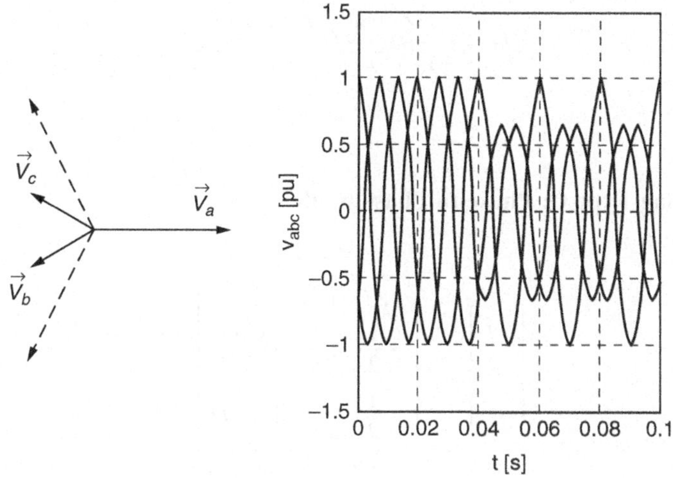

Fig. 4.11 Sag type C [44]

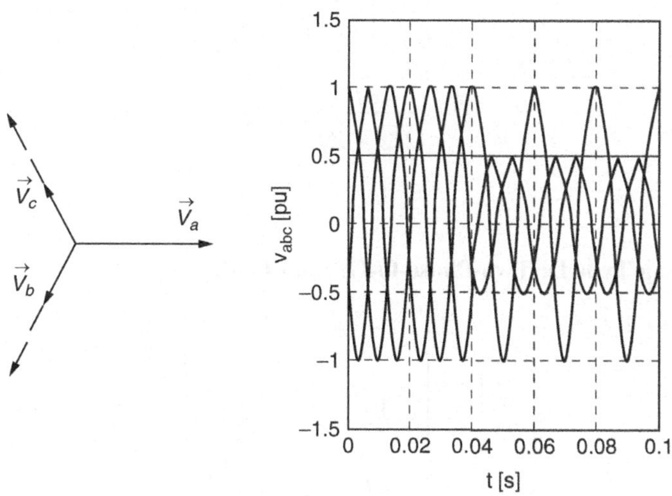

Fig. 4.12 Sag type D [44]

$$\begin{bmatrix} \overrightarrow{\text{va+}} \\ \overrightarrow{\text{va}-} \\ \overrightarrow{\text{va0}} \end{bmatrix} = \begin{bmatrix} \frac{1}{3}\left(2+\vec{D}\right) \\ -\frac{1}{3}\left(1-\vec{D}\right) \\ -\frac{1}{3}\left(1-\vec{D}\right) \end{bmatrix} \overrightarrow{Vsa^{+}} \tag{4.15}$$

$$\vec{D} = \frac{\overrightarrow{Zf^+} + \overrightarrow{Zf^-} + \overrightarrow{Zf^0}}{3\overrightarrow{Zs} + \overrightarrow{Zf^+} + \overrightarrow{Zf^-} + \overrightarrow{Zf^0}} \tag{4.16}$$

4.4.1.3 Sag Type C: Phase-to-Phase Fault

$$\begin{bmatrix} \overrightarrow{va} \\ \overrightarrow{vb} \\ \overrightarrow{vc} \end{bmatrix} = \begin{bmatrix} . & 1 & . \\ -\frac{1}{2} & . & -j\frac{\sqrt{3}}{2}\vec{D} \\ -\frac{1}{2} & . & +j\frac{\sqrt{3}}{2}\vec{D} \end{bmatrix} \overrightarrow{Vsa^+} \tag{4.17}$$

$$\begin{bmatrix} \overrightarrow{va+} \\ \overrightarrow{va-} \\ \overrightarrow{va0} \end{bmatrix} = \begin{bmatrix} \frac{1}{2}\left(1 + \vec{D}\right) \\ \frac{1}{2}\left(1 - \vec{D}\right) \\ 0 \end{bmatrix} \overrightarrow{Vsa^+} \tag{4.18}$$

$$\vec{D} = \frac{\overrightarrow{Zf^+} + \overrightarrow{Zf^-}}{2\overrightarrow{Zs} + \overrightarrow{Zf^+} + \overrightarrow{Zf^-}} \tag{4.19}$$

4.4.1.4 Sag Type D: Two-Phase-to-Ground Fault

$$\begin{bmatrix} \overrightarrow{va} \\ \overrightarrow{vb} \\ \overrightarrow{vc} \end{bmatrix} = \begin{bmatrix} . & 1 & . \\ -\frac{1}{2}\vec{D} & . & -j\frac{\sqrt{3}}{2}\vec{D} \\ -\frac{1}{2}\vec{D} & . & +j\frac{\sqrt{3}}{2}\vec{D} \end{bmatrix} \overrightarrow{Vsa^+} \tag{4.20}$$

$$\begin{bmatrix} \overrightarrow{va+} \\ \overrightarrow{va-} \\ \overrightarrow{va0} \end{bmatrix} = \begin{bmatrix} \frac{1}{3}\left(1 + 2\vec{D}\right) \\ \frac{1}{3}\left(1 - \vec{D}\right) \\ \frac{1}{3}\left(1 - \vec{D}\right) \end{bmatrix} \overrightarrow{Vsa^+} \tag{4.21}$$

$$\vec{D} = \frac{\overrightarrow{Zf}}{Zs + \overrightarrow{Zf}} \tag{4.22}$$

4.5 Control of Grid Converters Under Grid Faults

The electrical network is a dynamical system whose behavior depends upon many factors, as, for instance, constraints set by power generation systems, the occurrence of grid faults and other contingences, the excitation of resonances, or the existence of nonlinear loads. Therefore, grid-connected power converters should be designed bearing in mind that they should guarantee a proper operation under generic grid voltage conditions, being especially important to design control algorithms that ensure a robust and safe performance under abnormal grid conditions [44, 52].

The ever-increasing integration of distributed generation systems, which should fulfil the tight requirements imposed by the grid operator, mainly regarding low-voltage ride through and grid support during transient grid faults, has encouraged researchers to improve the conventional control solutions for grid-connected power converters. Despite the fact that the control of power converters under abnormal grid conditions is not a new issue, most of the studies within this field were mainly focused on the control of active rectifiers. The main concern in such applications was to guarantee a proper performance on the DC side of the converter under grid faults. In the grid connection of distributed generators, the interaction between the power converter and the networks under balanced and unbalanced faults is a crucial matter, since it is necessary not only to guarantee that any protection of the converter would not trip but also to support the grid voltage under such faulty operating conditions.

Occurrences of grid faults usually give rise to the appearance of unbalanced grid voltages at the point of connection of the power converter. Under unbalanced conditions, the currents injected into the grid lose their sinusoidal and balanced appearance. The interaction between such currents and the unbalanced grid voltages may give rise to uncontrolled oscillations in the active and reactive power delivered to the network. The proper operation of the power converter under such conditions is a challenging control issue. However, as will be shown in this part, the injection of such unbalanced currents may also give rise to other useful effects. For instance, the injection of a proper set of unbalanced currents under unbalanced grid voltage conditions allows attenuating power oscillations, maximizing the instantaneous power delivery, or balancing the grid voltage at the point of connection. However, the injection of unbalanced currents into the grid cannot be accurately achieved by using most of the conventional current controllers currently implemented in the industry. For this reason, some improved control structures specifically designed to inject unbalanced currents into the grid will be presented in this part [47].

Depending on the objective of the control strategy used to generate the reference currents during grid faults, the overall performance of the power converter and its interaction with the electrical grid will vary considerably. Therefore, reference current generation under grid faults is another crucial issue in the control of power converters. Since different current set points can be found to deliver a given amount of active and reactive power, a detailed study about implementation of different reference current generation strategies, together with an analysis of their performance under generic conditions, will be presented in this part. The design of such strategies will be performed considering not only the shape of the currents but also the behavior of the instantaneous power delivered by the converter. The currents injected by the power converter into the phases of the grid should always be under control, even though the grid voltage experiences strong variations. Therefore, the control algorithms setting the reference currents should estimate the instantaneous performance of these phase currents at any time, even during transient faults, in order to avoid any overcurrent tripping. For this reason, the last but not the least important issue tackled in this part is related to the calculation of the maximum power that can be delivered to the grid, without overpassing the current limits of the power converter. The control of grid-connected power converters under grid faults studied in this part represents an essential complement to the methods treated previously regarding the control of power converters under balanced grid voltage conditions.

4.5.1 Sequence Extraction

In steady state, voltages and currents can be decomposed into positive, negative, and zero sequences. In this article, AC side zero sequence is discarded due to natural filtering provided by D/Yn transformer between the VSC and the grid. If positive "+" or negative "−" Park transformation is applied to a system that contains positive and negative sequences, a combination of a constant term and a term oscillating at double-line frequency (2ω) is obtained:

$$\begin{bmatrix} x_d{}^+ \\ x_q{}^+ \end{bmatrix} = x^+ \begin{bmatrix} \cos(\theta^+ - \alpha) \\ \sin(\theta^+ - \alpha) \end{bmatrix} + x^- \begin{bmatrix} \cos(-2\omega t - \theta^- - \alpha) \\ \sin(-2\omega t - \theta^- - \alpha) \end{bmatrix} \tag{4.23}$$

$$\begin{bmatrix} x_d{}^- \\ x_q{}^- \end{bmatrix} = x^- \begin{bmatrix} \cos(\theta^- - \alpha) \\ -\sin(\theta^- - \alpha) \end{bmatrix} + x^+ \begin{bmatrix} \cos(2\omega t + \theta^+ - \alpha) \\ \sin(2\omega t + \theta^+ - \alpha) \end{bmatrix} \tag{4.24}$$

where $x_d{}^+$, $x_q{}^+$, $x_d{}^-$, and $x_q{}^-$ are voltages or currents in positive and negative *dq* frames; x^+ and x^- denote amplitudes of positive and negative sequences, respectively; $\Theta^+$and Θ^- are phase shifts of positive and negative sequence components; and α is the phase shift of Park transformation. If the oscillating terms are filtered with a low-pass or band-stop filter, the phase margin of the system is reduced considerably [44]; therefore, the above equations are not used directly but rather as a basis for more efficient techniques.

4.5.2 Decoupling by Compensation

This method is based on the fact that the amplitude of the oscillations x^- in (4.23) is equal to x^- in (4.24). A similar observation is valid for x^+. It is thus possible to eliminate the oscillations by subtracting them [47]. A combination of a notch filter (NF) and double-line frequency Park transforms is used to produce an oscillating signal, which is then subtracted; these 2ω oscillations on the synchronous reference frames should be canceled out in order to achieve full control of the injected currents under unbalanced conditions. In the following, some solutions oriented to overcome this drawback are presented. Maybe the most straightforward solution to attenuate the effects of the 2ω oscillations consists of filtering the measured currents by using a notch filter (NF), tuned at 2ω as shown in Fig. 4.13. The blocks P represent the Park transformations, the reference values for currents in *dq* frames are supposed to be constant, they can be taken as feedforward terms, and in this case, filtering should only be applied to the error signal between measured and reference currents [44]. Such approach has similar performance.

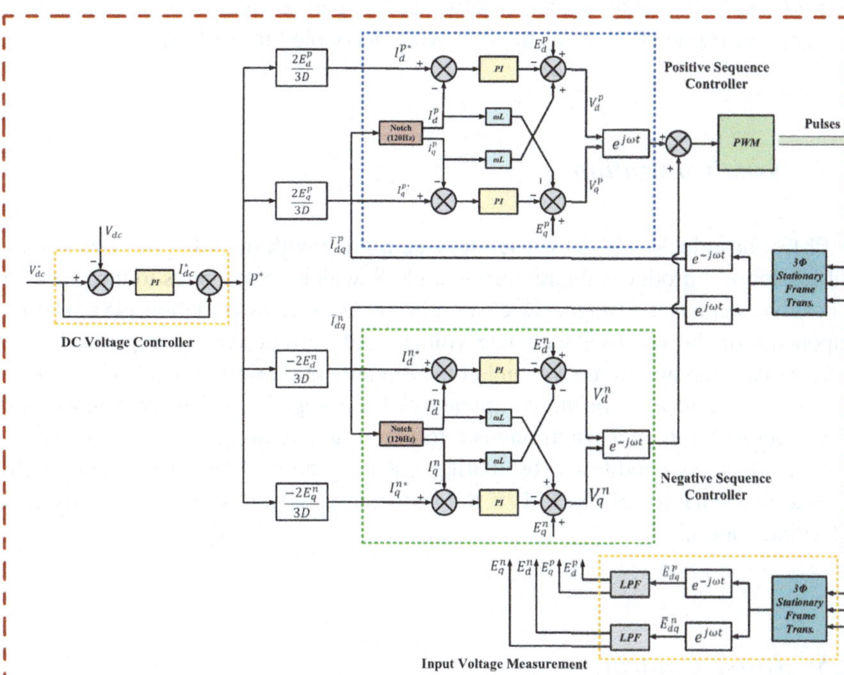

Fig. 4.13 Sequence decoupling: (a) by notch filter and in *dq* frame

4.5.3 Decoupling in dq Frame

It is possible to decouple the sequences in the stationary *dq* frame, knowing that in a sinusoidal steady state, the following equations are true for positive and negative sequence currents:

$$\begin{aligned} idq^{+} &= \begin{bmatrix} id^{+} \\ iq^{+} \end{bmatrix} = \begin{bmatrix} -id^{+} \\ -iq^{+} \end{bmatrix} + \begin{bmatrix} \tilde{i}d^{+} \\ \tilde{i}q^{+} \end{bmatrix} = I^{+} \begin{bmatrix} \cos(\theta^{+} - \alpha) \\ \sin(\theta^{+} - \alpha) \end{bmatrix} \\ &+ I^{-} \cos(\theta^{-} - \alpha) \begin{bmatrix} \cos(2\omega t) \\ -\sin(2\omega t) \end{bmatrix} + I^{-} \sin(\theta^{-} - \alpha) \begin{bmatrix} \sin(2\omega t) \\ \cos(2\omega t) \end{bmatrix} \end{aligned} \tag{4.25}$$

$$\begin{aligned} idq^{-} &= \begin{bmatrix} id^{-} \\ iq^{-} \end{bmatrix} = \begin{bmatrix} -id^{-} \\ -iq^{-} \end{bmatrix} + \begin{bmatrix} \tilde{i}d^{-} \\ \tilde{i}q^{-} \end{bmatrix} = I^{-} \begin{bmatrix} \cos(\theta^{-} - \alpha) \\ \sin(\theta^{-} - \alpha) \end{bmatrix} \\ &+ I^{+} \cos(\theta^{+} - \alpha) \begin{bmatrix} \cos(2\omega t) \\ \sin(2\omega t) \end{bmatrix} + I^{+} \sin(\theta^{+} - \alpha) \begin{bmatrix} -\sin(2\omega t) \\ \cos(2\omega t) \end{bmatrix} \end{aligned} \tag{4.26}$$

where id^{+}, iq^{+}, id^{-}, and iq^{-} are positive and negative sequences of voltages or currents in *dq* frame. This gives the following decoupled terms [48]:

4.5.4 Synchronization

The PLL (phase-locked loop) is used to synchronize with the AC grid. Based on a feedback loop, it produces the reference angle θ, which corresponds to the phase of the positive sequence voltage, which is used for Park transformation [48]. The *dq* components of the positive sequence voltage and current vectors expressed on a synchronous reference frame rotating at the fundamental grid frequency ω, whereas vd^{-}, vq^{-}, id^{-}, and iq^{-} are the components of the negative sequence voltage and current vectors lying on a synchronous reference frame rotating at $-\omega$, respectively. Under unbalanced conditions, oscillating values produced by (4.25) and (4.26) influence the behavior of the PLL if the instantaneous values are fed directly into the feedback loop.

4.5.5 Inner Control

Figure 4.14 shows the schematic diagram of the grid-side converter. The dynamic equations in a synchronous frame rotating with the grid voltage are:

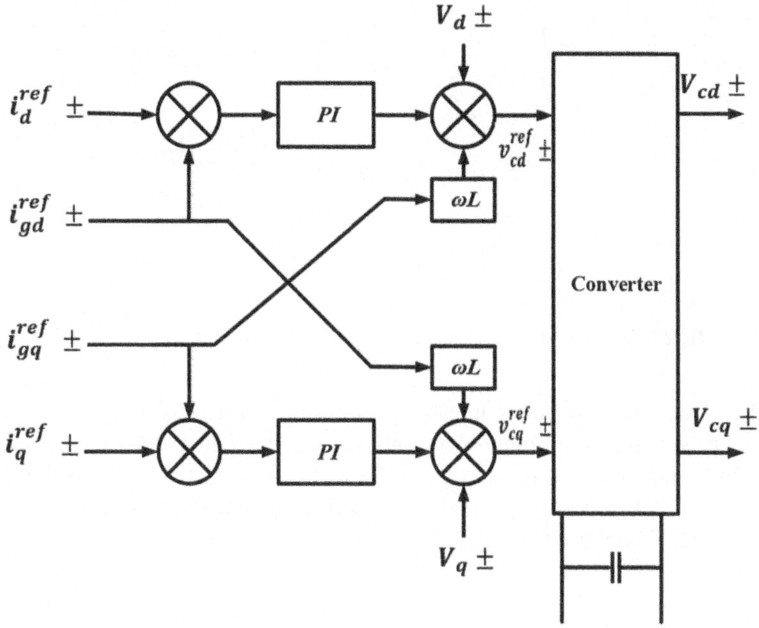

Fig. 4.14 Control of VSC station currents in positive and negative *dq* frames

$$V_{\mathrm{cd}} = \left(Ri_{\mathrm{d}} + L\frac{di_{\mathrm{d}}}{d_{\mathrm{t}}}\right) + \left(V_{\mathrm{gd}} - \omega Li_{\mathrm{q}}\right) \tag{4.27}$$

$$V_{\mathrm{cq}} = \left(Ri_{\mathrm{q}} + L\frac{di_{\mathrm{q}}}{d_{\mathrm{t}}}\right) + \left(V_{\mathrm{gq}} - \omega Li_{\mathrm{d}}\right) \tag{4.28}$$

PI controller is the most common type in *dq* frame due to its ability to track DC and low-frequency signals by combining the current system and grid, side dynamic.

4.5.6 *Control in the DQ Reference Frame*

To control positive and negative sequence currents, it is easy to use classic PI regulators in the rotating positive and negative *dq* frames with grid voltage and current feed forward to produce reference signals for the voltage grid side ((4.29) and (4.30)). The system for positive *dq* frame is indicated in Fig. 4.11, and the negative *dq* frame system is analogous.

$$\begin{bmatrix} ed^+ \\ eq^+ \end{bmatrix} = \begin{bmatrix} vd^+ \\ vq^+ \end{bmatrix} - \begin{bmatrix} \mathrm{PI}(s)\left(id^{+\mathrm{ref}} - id^+\right) \\ \mathrm{PI}(s)\left(iq^{+\mathrm{ref}} - iq^+\right) \end{bmatrix} - \omega\left(\frac{L_\mathrm{p}}{2} + L_\mathrm{g}\right)\begin{bmatrix} -iq^+ \\ id^+ \end{bmatrix} \tag{4.29}$$

$$\begin{bmatrix} ed^- \\ eq^- \end{bmatrix} = \begin{bmatrix} vd^- \\ vq^- \end{bmatrix} - \begin{bmatrix} \mathrm{PI}(s)\left(id^{-\mathrm{ref}} - id^-\right) \\ \mathrm{PI}(s)\left(iq^{-\mathrm{ref}} - iq^-\right) \end{bmatrix} - \omega\left(\frac{L_\mathrm{p}}{2} + L_\mathrm{g}\right)\begin{bmatrix} iq^- \\ -id^- \end{bmatrix} \tag{4.30}$$

4.5.7 *Outer Control*

In HVDC applications, the outer loop controls active/reactive power and DC voltage, producing the reference currents for the inner loop. Power regulation is based on an integral control with gains K_{ip} and K_{IQ} for active and reactive power (4.31, 4.32). If balanced conditions and perfect synchronization are considered, (4.31) and (4.32) are used:

$$id^{\mathrm{ref}} = \frac{1}{vd}\frac{K_{\mathrm{ip}}}{s}\left(P_{\mathrm{ac}}{}^{\mathrm{ref}} - P_{\mathrm{ac}}\right) \tag{4.31}$$

$$iq^{\mathrm{ref}} = \frac{1}{vd}\frac{K_{\mathrm{iQ}}}{s}\left(Q_{\mathrm{ac}}{}^{\mathrm{ref}} - Q_{\mathrm{ac}}\right) \tag{4.32}$$

As for the DC voltage, a second-order closed-loop transfer function is typically used, based on the control as shown in Eq. (4.33):

$$id^{\mathrm{ref}} = \left(\frac{Ki_{\mathrm{vdc}}}{s} + Kp_{\mathrm{vdc}}\right)\left(\mathrm{vdc}^{\mathrm{ref}} - \mathrm{vdc}\right) \tag{4.33}$$

The DC side Idc current is considered as a constant disturbance that can be rejected with this regulator and therefore is not included in the control equation.

4.5.8 *Current Reference Distribution*

The signals produced by the outer control loop need to be converted to a set of reference currents for the internal loop. Depending on the objective during unbalance, it can be done using different techniques [47, 48], described in the following. The methods discussed in this paper are based on the following matrix equation for active/reactive powers, voltages, and currents at PCC in *dq* frame values:

$$\begin{bmatrix} -\overline{P} \\ -\overline{Q} \\ P_{c2} \\ P_{s2} \\ Q_{c2} \\ Q_{s2} \end{bmatrix} = \begin{bmatrix} vd^{+} & vq^{+} & vd^{-} & vq^{-} \\ vq^{+} & -vd^{+} & vq^{-} & -vd^{-} \\ vd^{-} & vq^{-} & vd^{+} & vq^{+} \\ vq^{-} & -vd^{-} & -vq^{+} & vd^{+} \\ vq^{-} & -vd^{-} & vq^{+} & -vd^{+} \\ -vd^{-} & -vq^{-} & vd^{+} & vq^{+} \end{bmatrix} \begin{bmatrix} id^{+} \\ iq^{+} \\ id^{-} \\ iq^{-} \end{bmatrix} \tag{4.34}$$

where $\overline{P}$ and $\overline{Q}$ are the average values over one period and P_{c2}, P_{s2}, Q_{c2}, and Q_{s2} are the amplitudes of the second harmonic oscillatory terms in phase and in quadrature with positive sequence voltage.

4.6 PNSC (Positive Negative Sequence Control)

This approach allows suppressing active/reactive power ripple when the reference of reactive/active power is zero as shown in Fig. 4.15:

$$\begin{bmatrix} id^{+} \\ iq^{+} \\ id^{-} \\ iq^{-} \end{bmatrix} = \frac{1}{|v^{+}|^{2} - |v^{-}|^{2}} \begin{bmatrix} vd^{+} & vq^{+} \\ vq^{+} & -vd^{+} \\ -vd^{-} & vq^{-} \\ -vq^{-} & -vd^{-} \end{bmatrix} \begin{bmatrix} -P \\ -Q \end{bmatrix} \tag{4.35}$$

If the amplitude of positive sequence voltage is equal to the amplitude of the negative sequence voltage, which is the case during close phase-to-phase faults, division by zero occurs; therefore, (4.35) should be used with care in real applications.

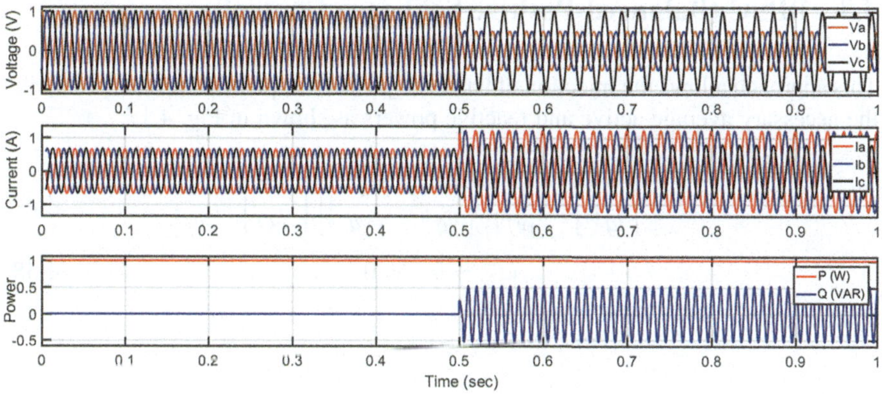

Fig. 4.15 Grid currents, active and reactive powers using PNSC control

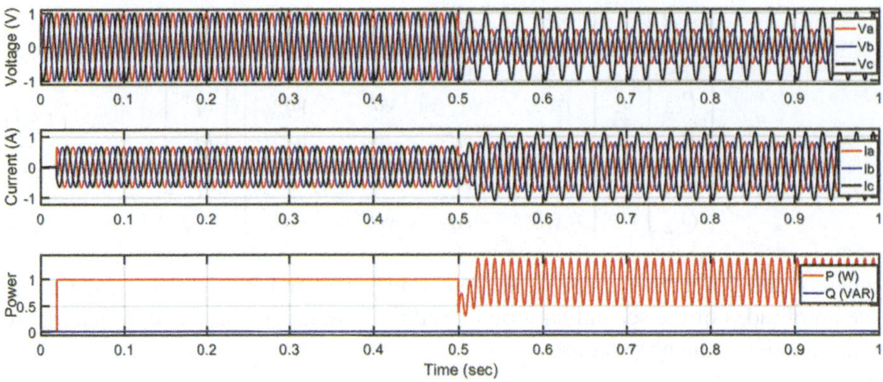

Fig. 4.16 Grid currents, active and reactive powers using AARC control

4.7 AARC (Average Active-Reactive Control)

This control method allows the injection of minimal RMS currents needed to produce the necessary average active and reactive power as shown in Fig. 4.16:

$$\begin{bmatrix} id^+ \\ iq^+ \\ id^- \\ iq^- \end{bmatrix} = \frac{1}{|v^+|^2 + |v^-|^2} \begin{bmatrix} vd^+ & vq^+ \\ vq^+ & -vd^+ \\ vd^- & -vq^- \\ vq^- & vd^- \end{bmatrix} \begin{bmatrix} -P \\ -Q \end{bmatrix} \tag{4.36}$$

4.8 BPSC (Balanced Positive Sequence Control)

This approach allows the injection of balanced positive sequence currents, producing the necessary average active and reactive powers as shown in Fig. 4.17:

$$\begin{bmatrix} id^+ \\ iq^+ \end{bmatrix} = \frac{1}{|v^+|^2} \begin{bmatrix} vd^+ & vq^+ \\ vq^+ & -vd^+ \end{bmatrix} \begin{bmatrix} -P \\ -Q \end{bmatrix} \tag{4.37}$$

$$id^- = iq^- = 0 \tag{4.38}$$

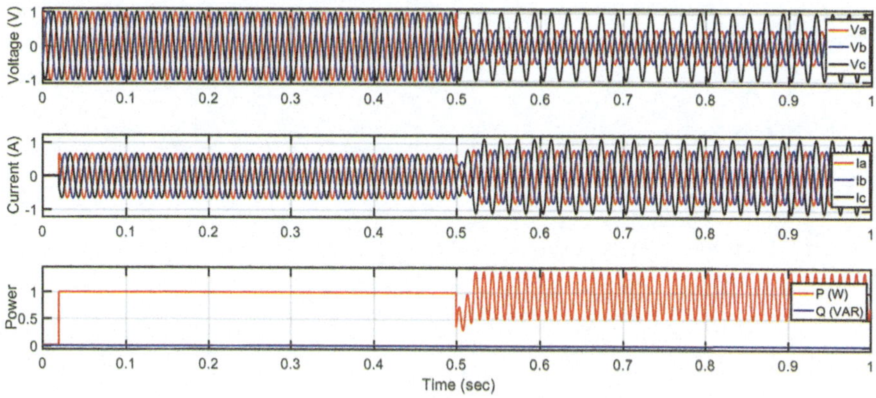

Fig. 4.17 Grid currents, active and reactive powers using BPSC control

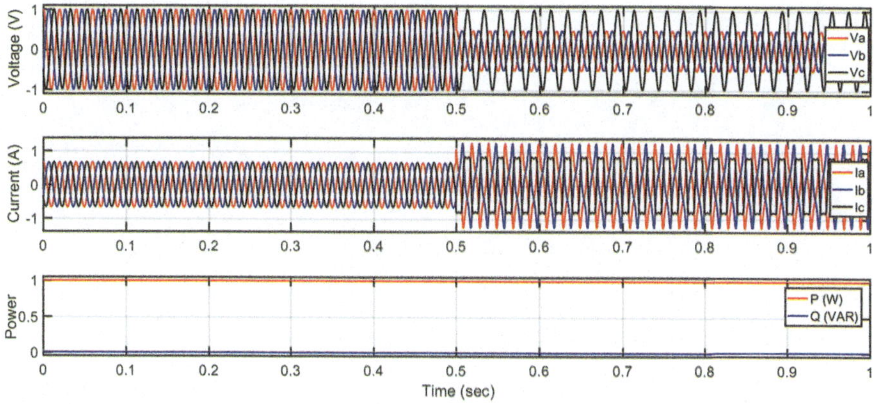

Fig. 4.18 Grid currents, active and reactive powers using IARC control

4.9 IARC (Instantaneous Active-Reactive Control)

This control method allows the injection of minimal RMS currents needed to produce the necessary average active and reactive power as shown in Fig. 4.18:

$$\begin{bmatrix} id^+ \\ iq^+ \\ id^- \\ iq^- \end{bmatrix} = \frac{1}{|v^+|^2 + |v^-|^2} \begin{bmatrix} vd^+ & vq^+ \\ vd^+ & -vd^+ \\ vd^- & -vq^- \\ vq^- & vd^- \end{bmatrix} \begin{bmatrix} -P \\ -Q \end{bmatrix} \tag{4.39}$$

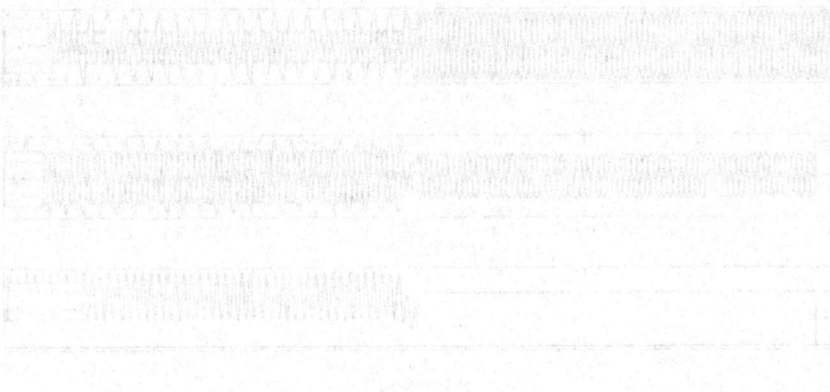

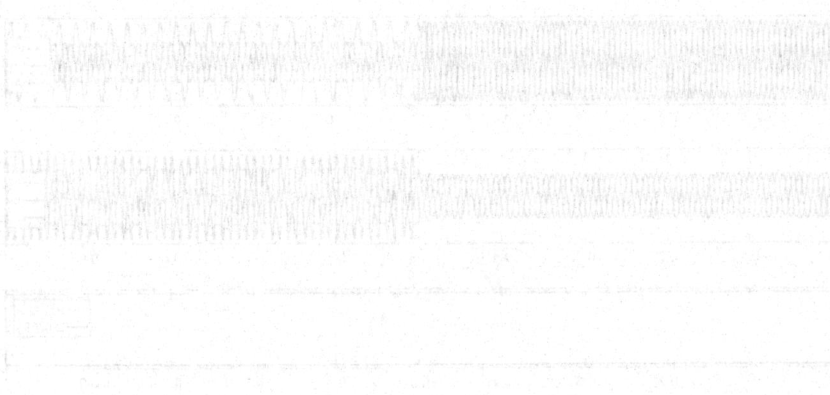

Chapter 5
VSC-HVDC Simulation Results

5.1 Introduction

The system considered in this book is given in Fig. 5.1. The HVDC system consists of two VSCs connected via a DC link. The control of converter VSC-1 is done by using DC voltage control, whereas the control of converter VSC-2 is done by active power control. In this chapter, the steady-state and transient behavior of control schemes in normal and fault conditions a VSC-based HVDC connected to a grid is studied. The mutual effect between the VSC-HVDC performance and its control scheme dynamics is investigated. The control scheme that is used is vector control and hysteresis current control, because it makes it possible to independently control the active power and reactive power. The different methods of calculating the reference current to control VSC behavior under different disturbance conditions are investigated. Different types of faults and protection scheme of VSC are also investigated.

5.2 Simulation Environment

The well-known MATLAB/"Simulink" is the used simulation environment for the VSC-based HVDC system as well as for the rest of the system elements, except the control schemes. Different fault conditions are initiated to cause transient conditions. The MATLAB outputs at the controlling points are investigated. The output of each control and different disturbance conditions is then analyzed (Table 5.1).

The simulation model in MATLAB/Simulink as given in Fig. 5.1 was tested for the controller performance under three different cases. In the MATLAB model, the direction of current is considered from AC system to the converter, and the power is considered positive when entering the converter. The test cases are presented as follows. All the test cases are plotted for time interval $t = 0$ s to $t = 0.5$ s.

N. F. Ibrahim, S. S. Dessouky, *Design and Implementation of Voltage Source Converters in HVDC Systems*, Power Systems,
https://doi.org/10.1007/978-3-030-51661-1_5

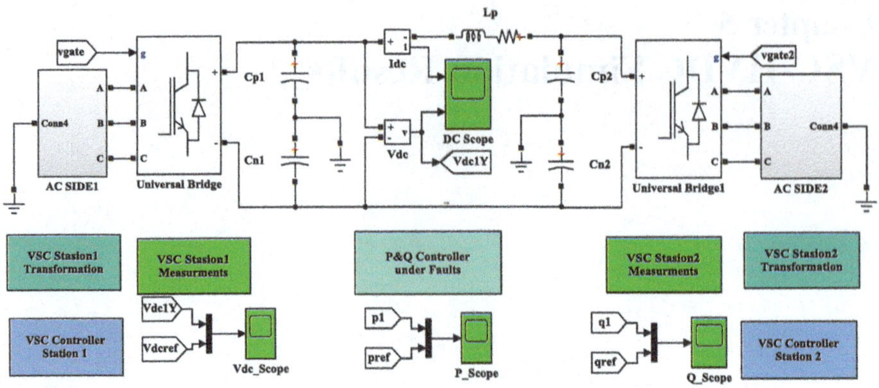

Fig. 5.1 Test system under consideration, MATLAB/Simulink

Table 5.1 Specifications of VSC used in the simulation

Parameters	Rating
Base frequency (ω_b)	314.1592 rad/sec
Frequency (f)	50 Hz
DC voltage (U_{DC})	700 V
AC voltage (V_n)	220Vrms
Rated power (P)	10 KW
Switching frequency	10 KHZ
DC capacitor	500 μF
Transmission length	75 km

5.2.1 Control Performance-Controller Tracking Capability

5.2.1.1 Control of DC Voltage

For simulating the transients in DC voltage, the operating DC voltage was first set to 600 V. At $t = 0.25$ s, the DC voltage reference is increased to 700 V for the initial validation of the control design such as control performance, control tracking capability, and response under a change in the reference signals (change in DC link voltage). As it is presented in Fig. 5.2, the DC-voltage has a good tracking for its reference.

5.2.1.2 Control of Active Power

In this case, the transients in active power, the active power was first set to 3 KW. At $t = 0.25$ s, the active power reference is increased to 5 KW for the initial validation of control design such as control performance, control tracking capability, and response under a change in the reference signals. This step is necessary for the

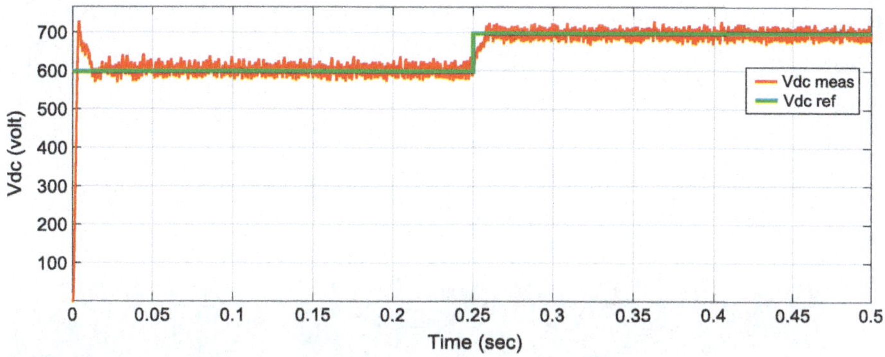

Fig. 5.2 DC voltage and control responses

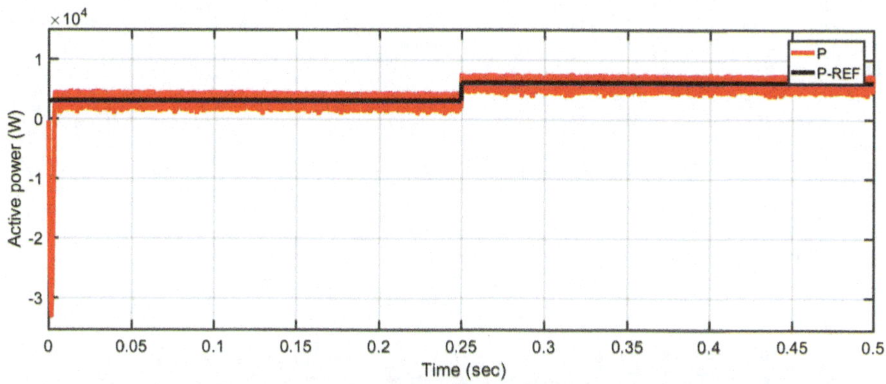

Fig. 5.3 Active power control response with different changes

initial validation of control design. As it is presented in Fig. 5.3, the active power has a good tracking for its reference. Because of the decoupling, almost no effect is observed in the control of active power.

5.2.2 Steady-State Simulation

Figure 5.4 presents the simulation results of voltage and current under steady-state operation at rectifier (VSC1) and inverter (VSC2) sides, the steady-state condition where the case VSC2 is transmitting power to VSC1, and the current and voltage waveforms for both VSC1 and VSC2. The results at the inverter side are almost similar to the rectifier.

The performance of the proposed control scheme, explained in this section, is tested for balance in grid voltages and currents. It can be seen from Fig. 5.4a, b that

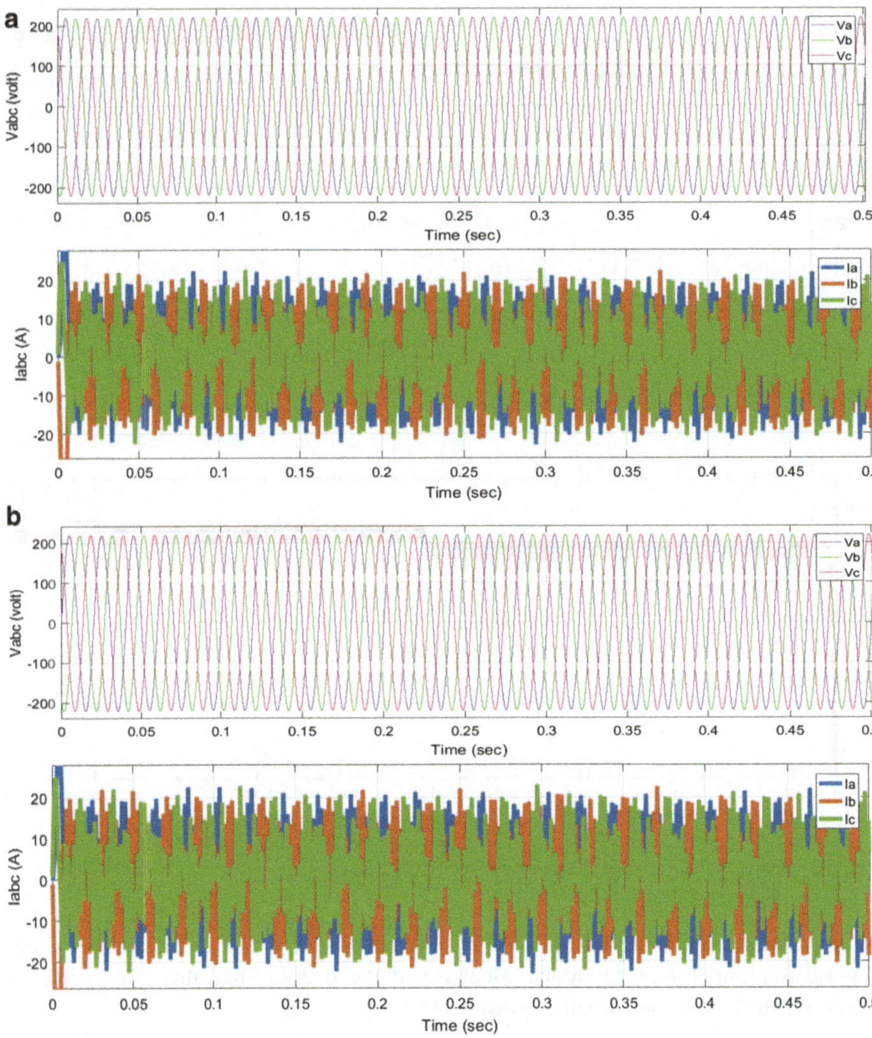

Fig. 5.4 Voltage and current waveforms for both VSC1 and VSC2

the peak value of the current is around 15 A. At all time, the amplitude value of the voltage is 220Vrms.

For the normal operation, the grid phase voltage and current are in the same phase angle, but some distortion harmonics are apparent in the current waveform (Fig. 5.5). As a result of these distortions, it is necessary to design the filter to reduce these distortions by analyzing harmonics.

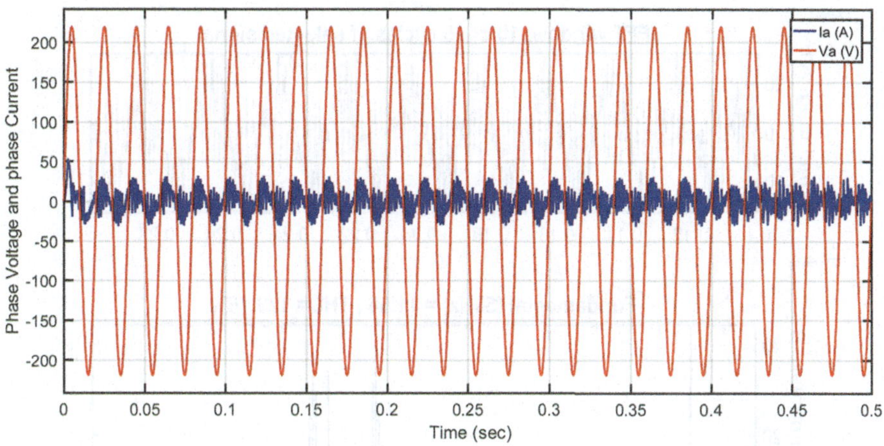

Fig. 5.5 The grid phase voltage and current

5.3 Harmonic Analysis

Under steady-state operation, the current and voltage waveforms are taken to evaluate the harmonic content of the designed VSC systems. The standard blocks in MATLAB for fast Fourier transform (FFT) and total harmonic distortion (THD) block are used to determine the harmonic magnitude of the line currents and voltages.

Figure 5.6 shows current and voltage waveforms at the point of common coupling (PCC) for phase-a in grid 1; the corresponding individual harmonics, harmonic distortion, and total harmonics distortion; and the FFT and THD for the line current and voltage after the filter. These measurements are consistent with the theoretical predictions from the PWM scheme. The switching frequency is 10 KHz for grid 1 and grid 2. In addition, a fifth harmonic component is also present, as observed in Fig. 5.6a, c and it can be seen that the percentage magnitude of the fifth harmonic component and THD value of grid current of phase-a is 123.15% but THD value of grid voltage of phase a is zero % as observed in Fig. 5.6b, d.

Figure 5.7 shows the harmonic spectrum and the waveform of the grid current of phase-a for the inverter and rectifier modes of operation, respectively. It can be seen that the percentage magnitude of the fifth harmonic component has reduced significantly compared to that in Fig. 5.6a, b. However, a small percentage of third and fifth harmonic components are present. In addition, the THD is reduced to 1.11% by designing the LCL filter.

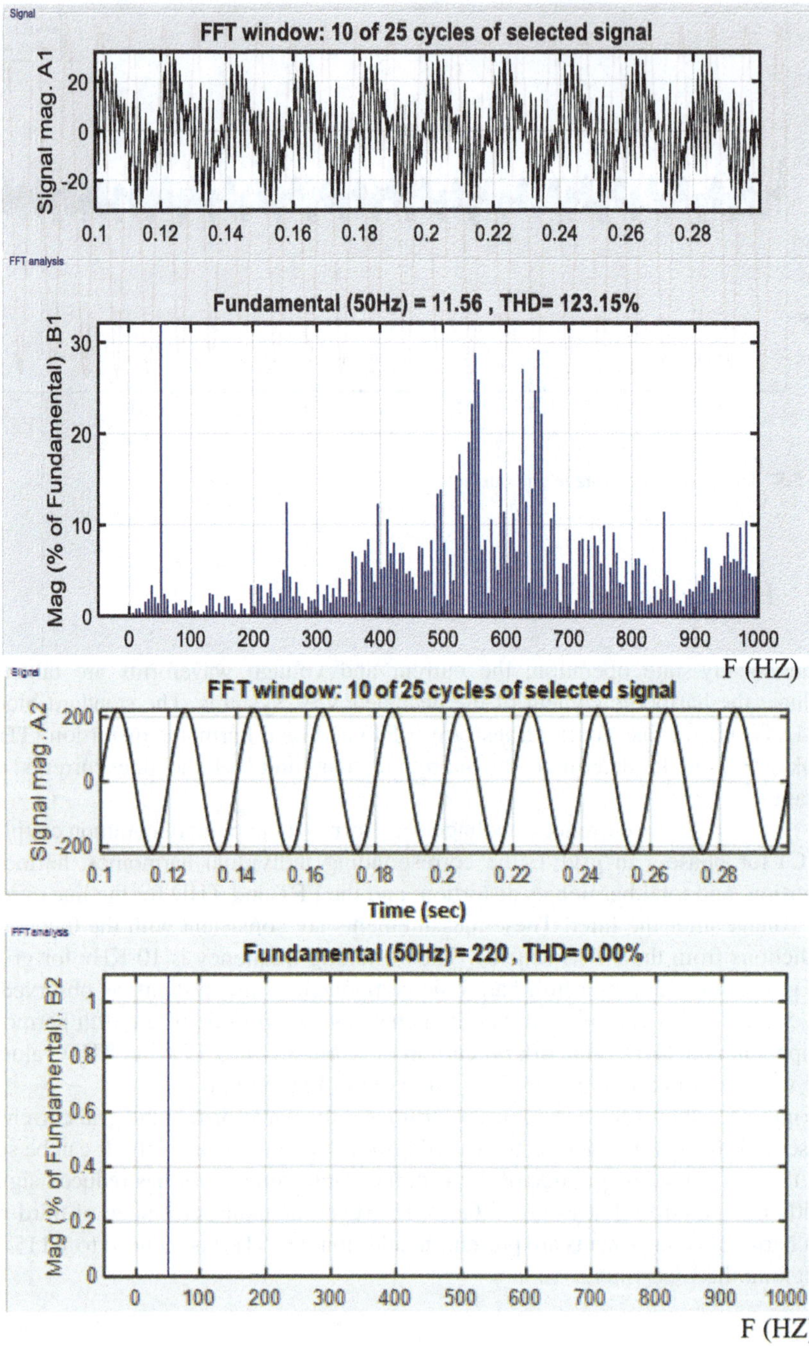

Fig. 5.6 Harmonic spectrum of injected current and voltage at converter terminal (before the filter), (**a**) individual harmonic distortion (%), (**b**) total harmonic distortion THD

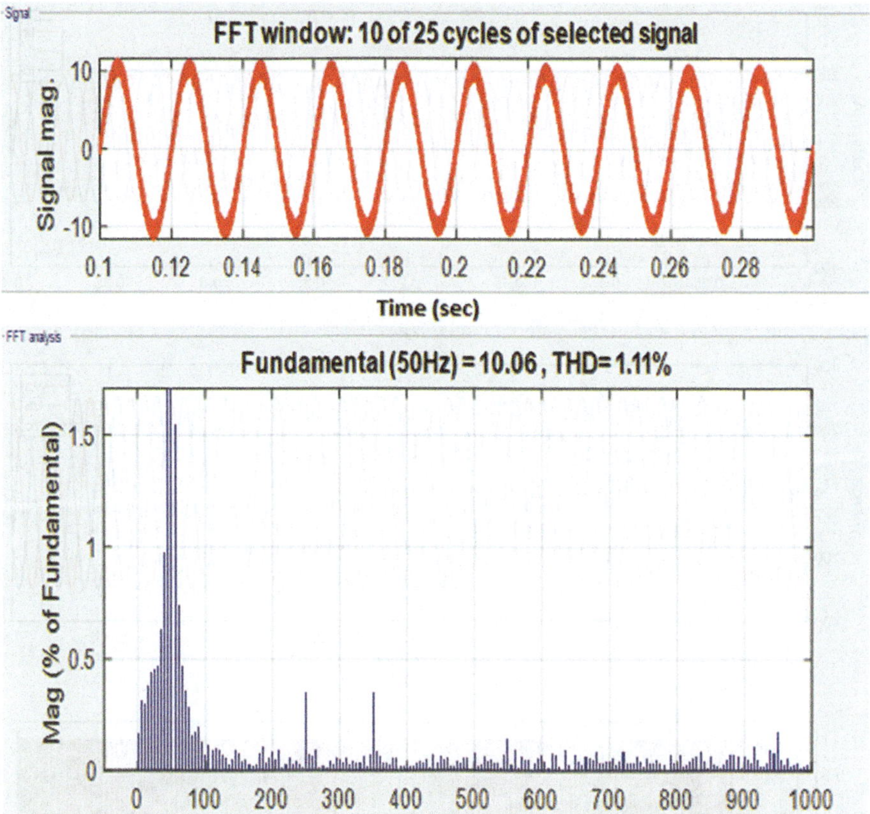

Fig. 5.7 Harmonic spectrum of injected current at converter terminal (after the filter), (**a**) individual harmonic distortion (%), (**b**) total harmonic distortion THD

5.4 Hysteresis Band Current Controller Simulation

In a steady-state condition, the simulation time is taken as $t = 0$ to $t = 0.4$ s. The waveforms in steady state for three phases are presented in Fig. 5.8. The three-phase supply voltage is shown in Fig. 5.8a. The three-phase load current is shown in Fig. 5.8b. The source current and voltage, from which we will get the knowledge of unity power factor, is shown in Fig. 5.8c. The response of hysteresis current control to phase a with the reference response is shown in Fig. 5.8e. The response of hysteresis current control to phase a with the reference response is shown in Fig. 5.8f under step change.

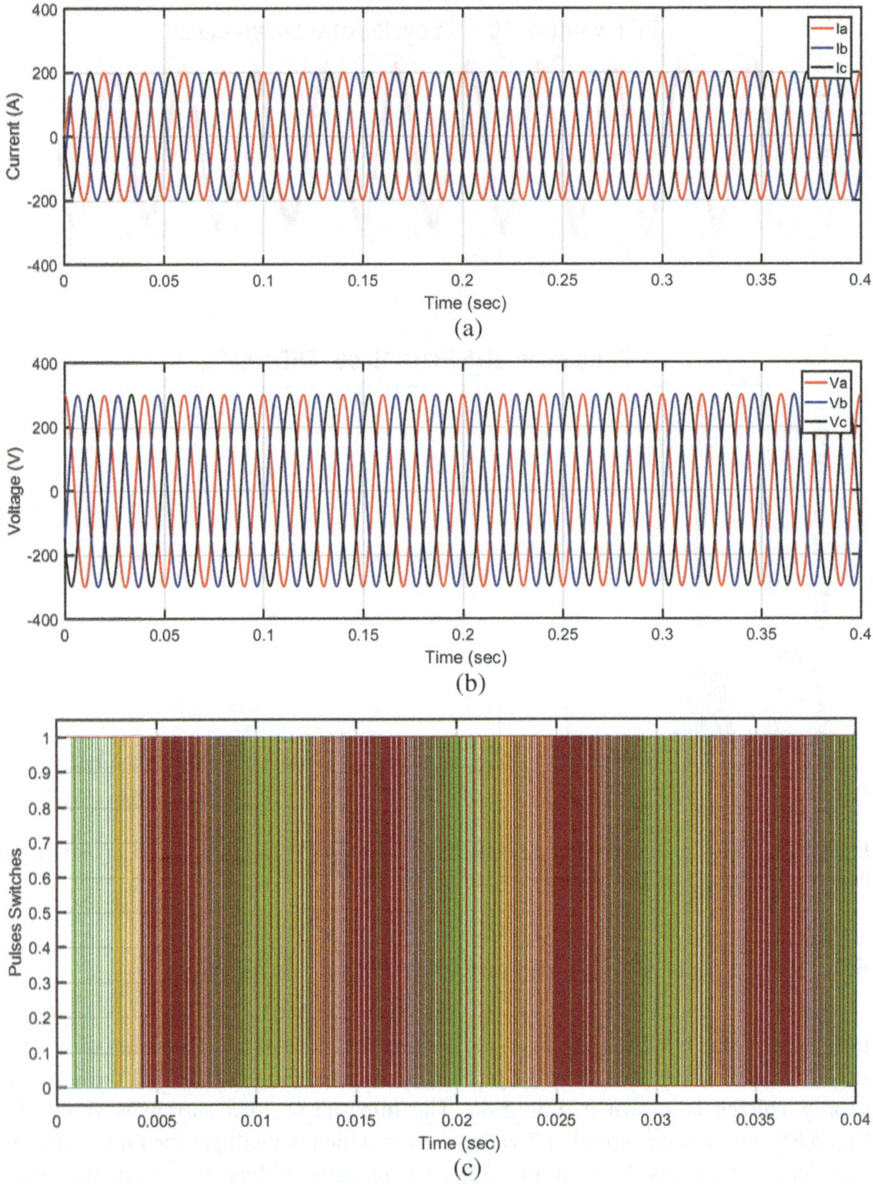

Fig. 5.8 The response of highest current controller method. (**a**) The three-phase load current. (**b**) The three-phase load voltage. (**c**) Gate pulses for inverter switches. (**d**) Current control responses. (**e**) Current control responses under change

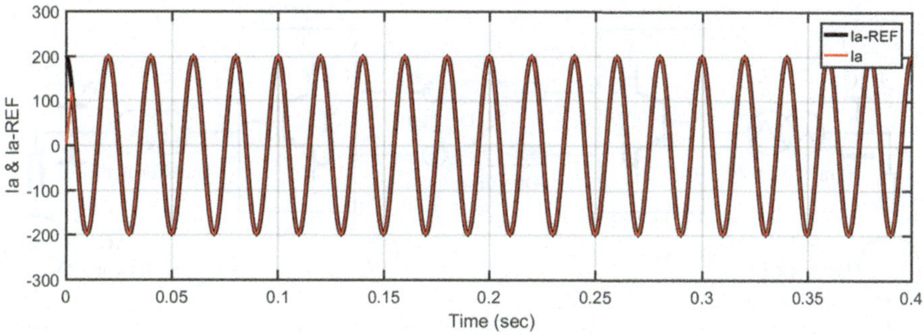

(d)

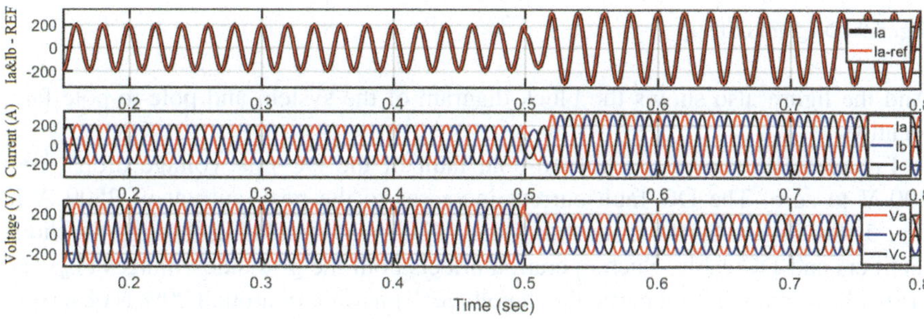

(e)

Fig. 5.8 (continued)

5.5 System Response Under Unbalance Fault Condition

5.5.1 DC Line Fault

This section focuses on the fault analysis of DC fault such as pole-to-pole, positive pole-to-ground, and negative pole-to-ground fault solution of VSC-HVDC system restoration after DC faults for overhead HVDC transmission line. MATLAB simulation software is used to simulate the VSC-based HVDC system. MATLAB is almost suitable for simulating the time domain instantaneous responses and is also popular for analyzing electromagnetic transients of electrical systems. MATLAB/ Simulink allows assembling the circuit, running the simulation, analyzing the results, and managing the data in a completely integrated graphical environment.

5.5.2 Pole-to-Pole Fault

In this case, the DC fault is located in the middle of the DC line and the fault duration is 0.4 s. The fault type permanent (free running response) can be seen in Fig. 5.9.

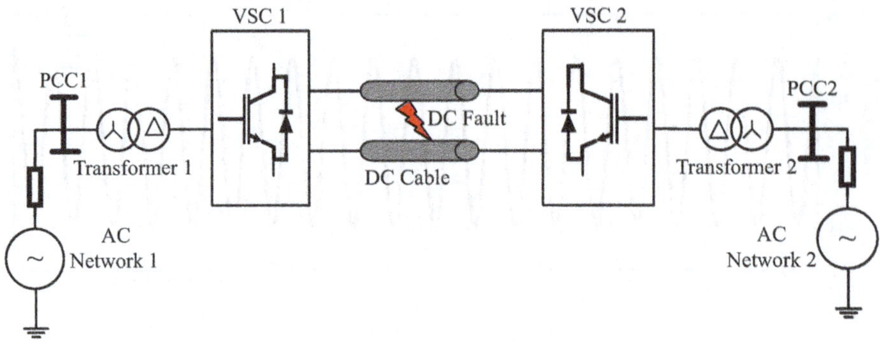

Fig. 5.9 Diagram of faults

And the figure also shows the block diagram of the system and pole-to-pole fault location.

Figure 5.10a shows the pole-to-pole fault of the DC-link voltage drops from 700 V to zero. The DC fault current increases to the peak current of 2500 A in Fig. 5.10b. Figure 5.10c, d illustrates the AC current circulating through the diode rectifiers and DC cable. Such operation affects both the grid-side voltages (V_g) and grid-side currents (I_g). In particular, a voltage depression of around 20% is observed, while the AC side current reaches an initial value of 300 A (during stage 1) and then falls to 150 A during the steady state (stage 3).

Figure 5.11 shows the diode currents will follow the waveform of the DC fault currents, but each diode will have a different transient current wave shape.

Figure 5.12 shows the DC side reactor (L_d) contributes to reducing the first peak of the diode and capacitor fault current. This reactor has no influence on the steady-state DC fault current. The value of this reactance is small typically below 5 mH per pole, in order to avoid dynamic interaction problems.

5.5.3 *Positive Pole-to-Ground Fault*

Positive pole-to-ground faults have a different impact on the system behavior than pole-to-pole faults. As mentioned before, the DC-link voltage does not drop to zero as shown in Fig. 5.13. The voltage drops from 700 V to grid voltage, but after a transient, it recovers back to an acceptable level. The fault type in this point is temporary.

Figure 5.13b shows the DC fault current increase to the peak current to 350 A. Figure 5.13c, d illustrates the AC current circulating through the diode rectifiers and DC cable. Such operation does not affect both the grid-side voltages (V_g) and grid-side currents (I_g)

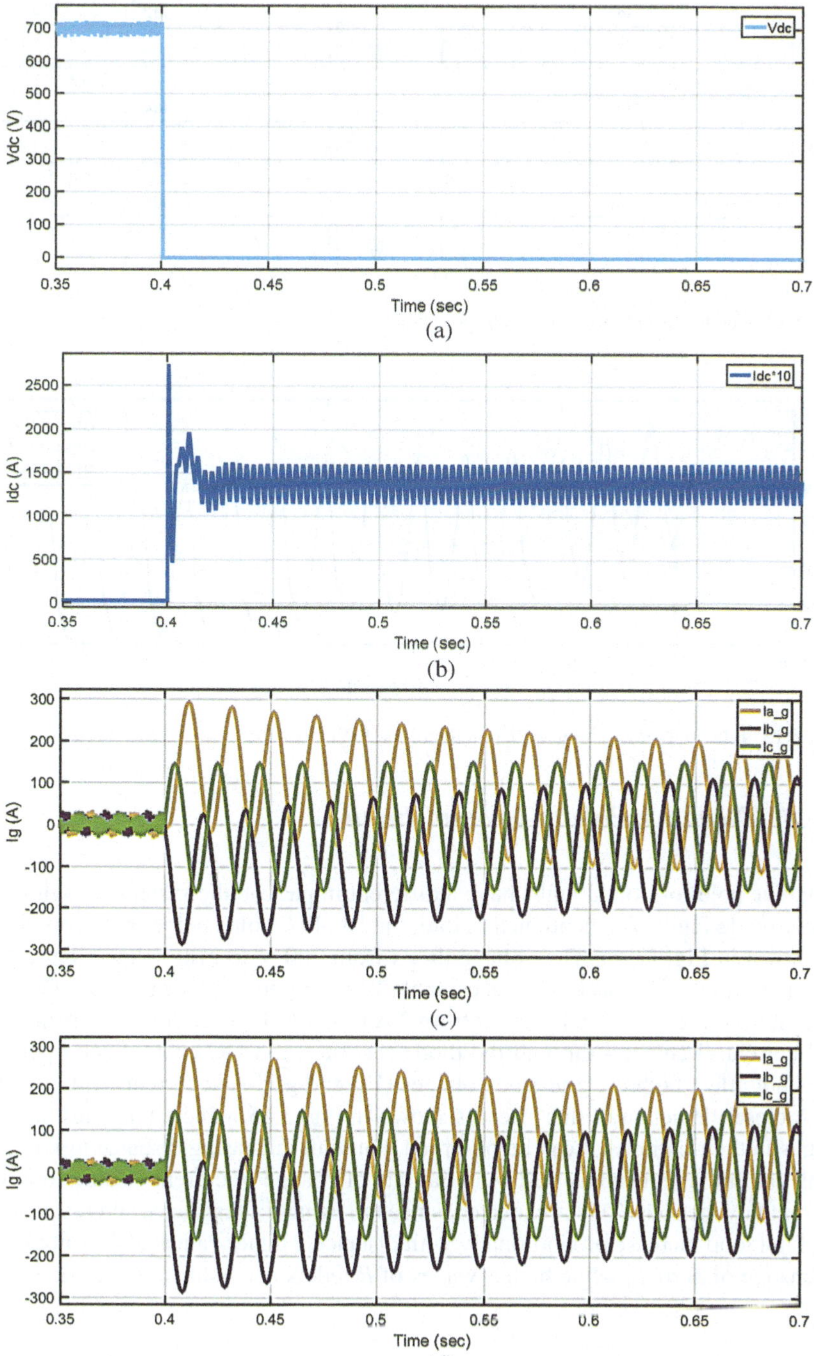

Fig. 5.10 VSC-HVDC cable pole-to-pole fault. (**a**) DC-link voltage. (**b**) DC-link current. (**c**) Grid-side currents. (**d**) Grid-side voltages

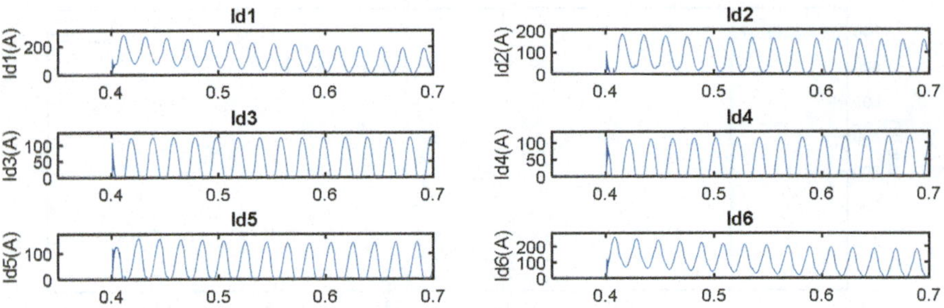

Fig. 5.11 Diode current through pole-to-pole fault

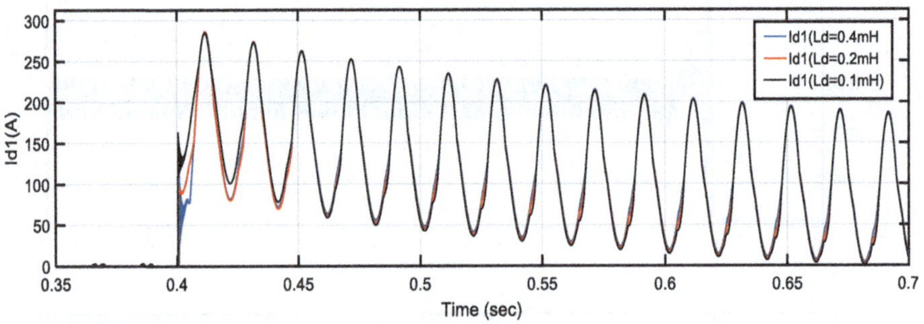

Fig. 5.12 Influence of DC inductor L_d on the diode fault current

5.5.4 Negative Pole-to-Ground Fault

Negative pole-to-ground faults have a different impact on the system behavior than pole-to-pole faults. As mentioned before, the DC-link voltage does not drop to zero as shown in Fig. 5.14a. The voltage drops from 700 V to grid voltage, but after a transient, it recovers back to an acceptable level. Figure 5.14b shows the DC fault current decrease to the peak current from 3 A to 1.5 A. Figure 5.14c, d illustrates the AC current circulating through the diode rectifiers and DC cable. Such operation does not affect both the grid-side voltages (V_g) and grid-side currents (I_g).

A closer depiction of this effect is shown in Fig. 5.15, which illustrates the fault stages in their entirety. It can be observed that with increasing distance to fault, the peak current is reduced as well as initial rate of change of current leading to longer current flow periods. This is expected, since the values of resistance and reactance in the fault loop increase proportionally with distance. Higher values of L limit the rate of change of current, while higher values of R reduce the value of the peak.

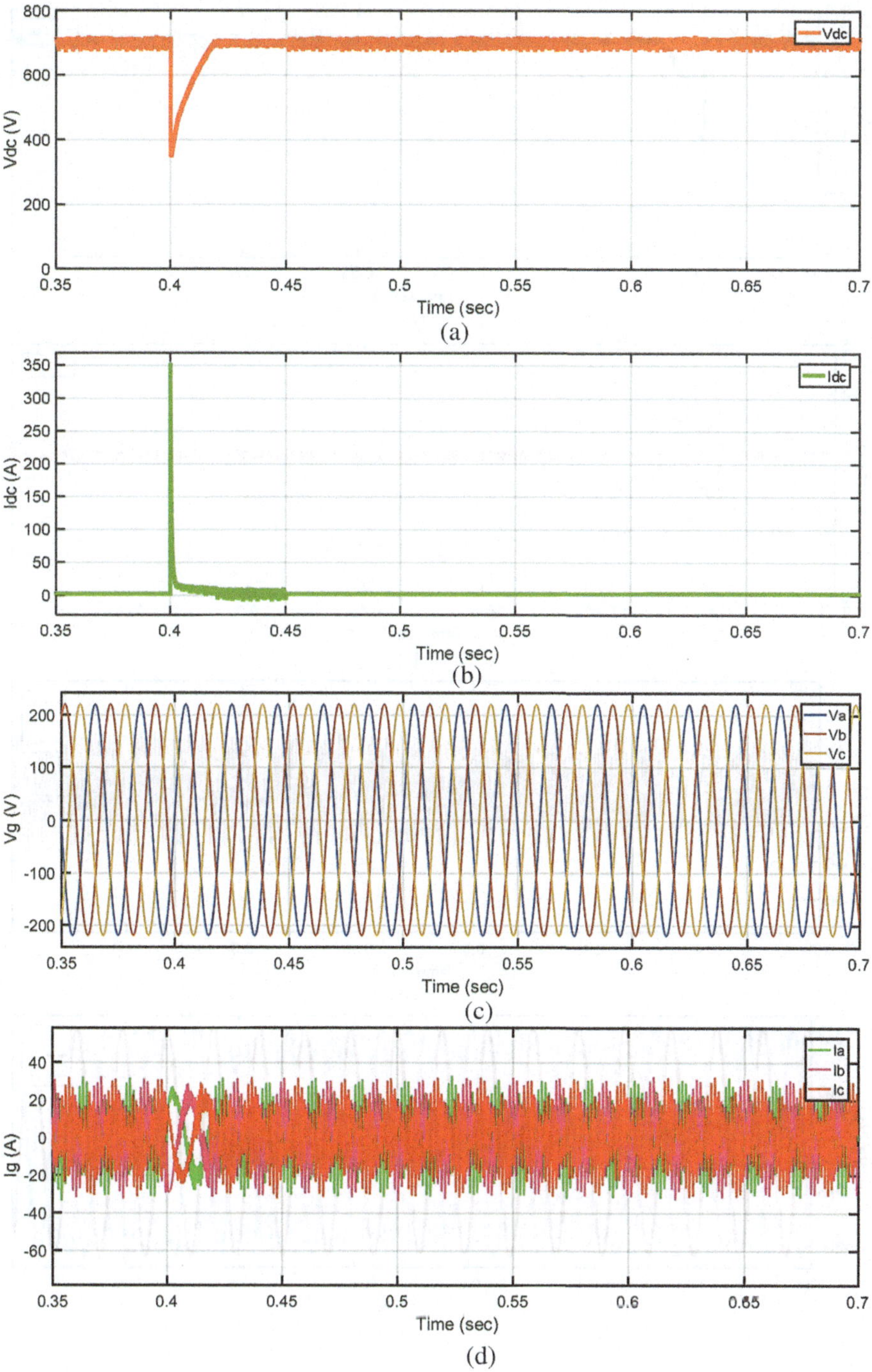

Fig. 5.13 VSC-HVDC cable positive pole-to-ground fault. (**a**) DC-link voltage. (**b**) DC-link current. (**c**) Grid-side voltages. (**d**) Grid-side currents

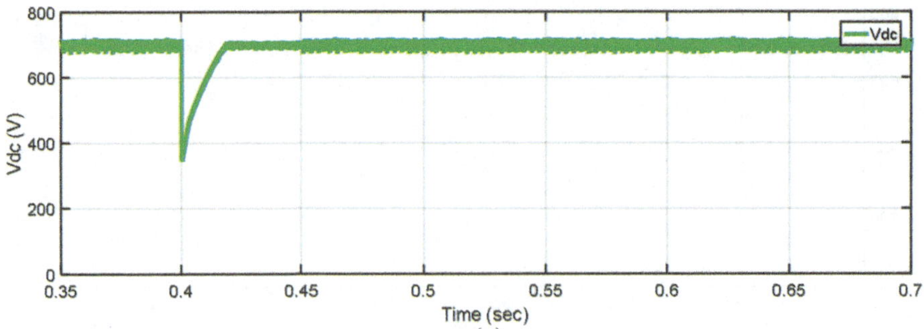

(a)

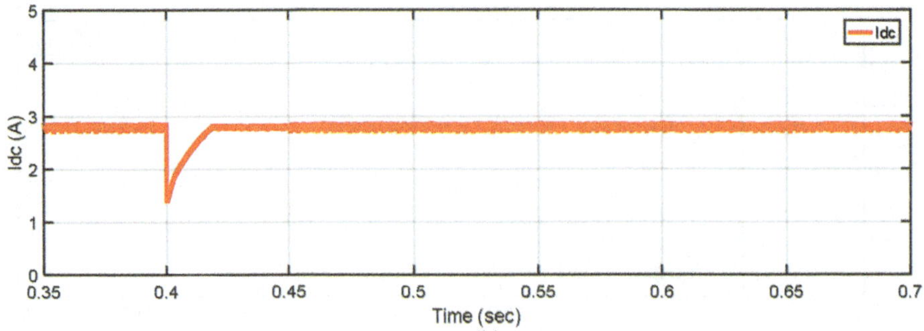

(b)

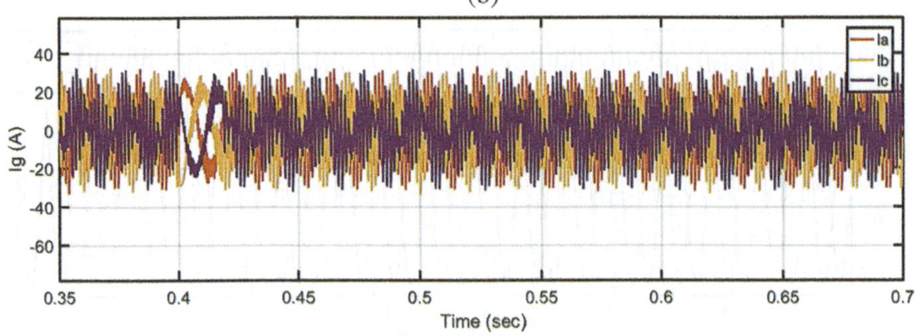

(c)

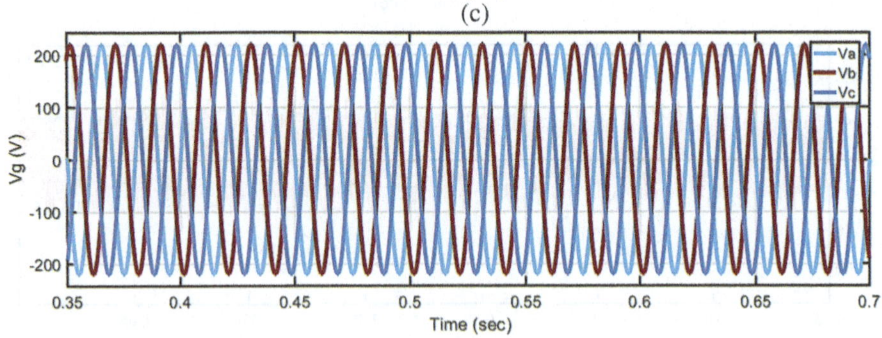

(d)

Fig. 5.14 VSC-HVDC cable negative pole-to-ground fault. (**a**) DC-link voltage. (**b**) DC-link current. (**c**) Grid-side currents. (**d**) Grid-side voltages

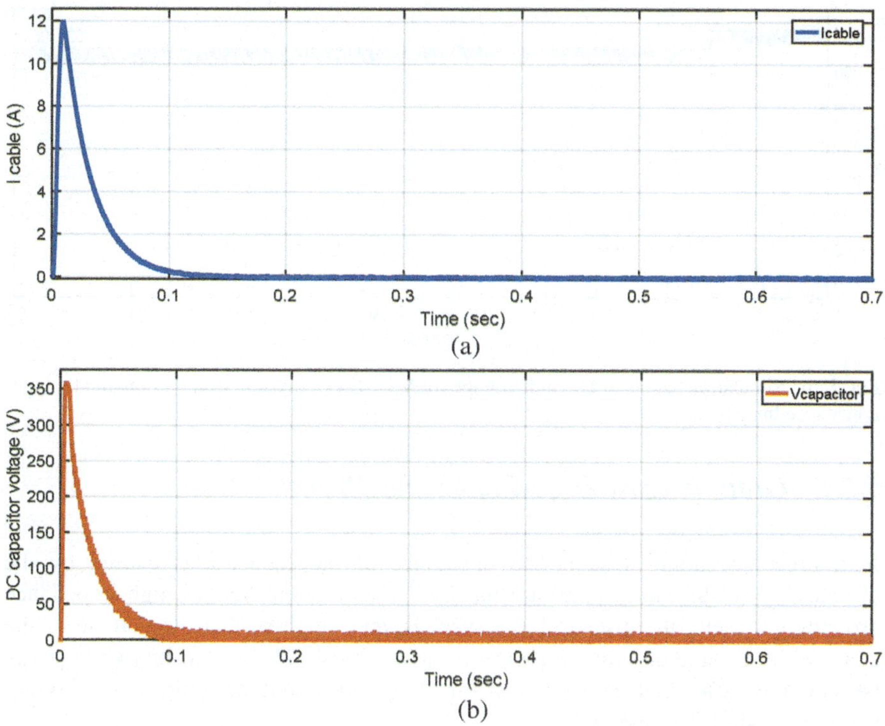

Fig. 5.15 Stage one illustration for pole-to-ground fault. (**a**) Free response of cable current for the first stage of pole-to-ground fault. (**b**) Free response of DC voltage for the first stage of pole-to-ground fault

5.6 System Response Under Recovery of DC Fault

The demand for transmission line faults is also to correctly detect the faults and then block and trip the converter for repairing. To avoid the overcurrent stress of freewheeling diodes after blocking, DC breakers can be introduced, for overhead line, temporary line-to-ground faults. The rebalance of capacitors can be realized by:

- Grounding by high-impedance branch
- Reconnecting transformer secondary winding to Yn type
- Monopolar scheme operation
- The design philosophy of fast DC breakers

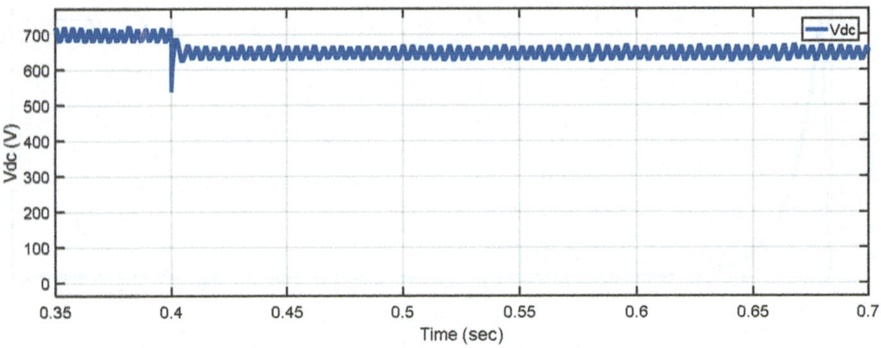

Fig. 5.16 Simulation results of DC-link voltage for VSC-HVDC system using grounding by high-impedance branch

5.6.1 Grounding by High-Impedance Branch

In indirect grounding system, the unbalance of the capacitor is caused by the discharging of the faulty pole and the DC voltage controller. In high-impedance grounding system, the discharging current is very small so that the voltage of the capacitor can maintain without any overcurrent stress. The line-to-line voltage can be recovered after fault as shown in Fig. 5.9, and output waveforms of recovery method are shown in Fig. 5.16.

5.7 Backup Protection of Converter by Design AC Breakers to Trip Grid

The basic demand of AC breakers is to isolate the grid from converter to avoid the AC system current feed the converter during DC fault as shown in Fig. 5.17. By using IGBT with an antiparalleled diode, the normal operating current can pass through IGBT of the sending end and diode of the receiving end. While the DC line is under a fault condition, the discharging current will force the IGBT to turn off immediately to realize the pole isolation. This can avoid AC breaker tripping, and then the converter can restart faster.

The simulation results are shown in Figs. 5.17 and 5.18. Due to the large current flowing moment of the fault which can be seen at the period of 0.4 s (Fig. 5.18b), while there is no visible effect of the circuit breaker at the relay operation. Figure 5.17 shows the effects of the relay when it is set at 125–135% of rated current of the DC current. The circuit breaker is disconnected in the system, and the current relay is activated. Control can be gained over circuit breaker if $K = 0$ or $K = 1$. The simulation results are shown in Fig. 5.18.

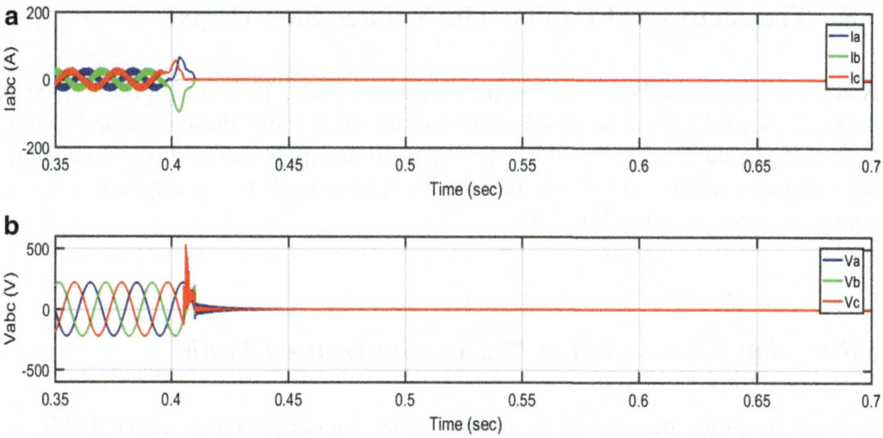

Fig. 5.17 Current and voltage waveform when DC fault occurs during circuit breaker tripping. (a) Grid-side current. (b) Grid-side voltage

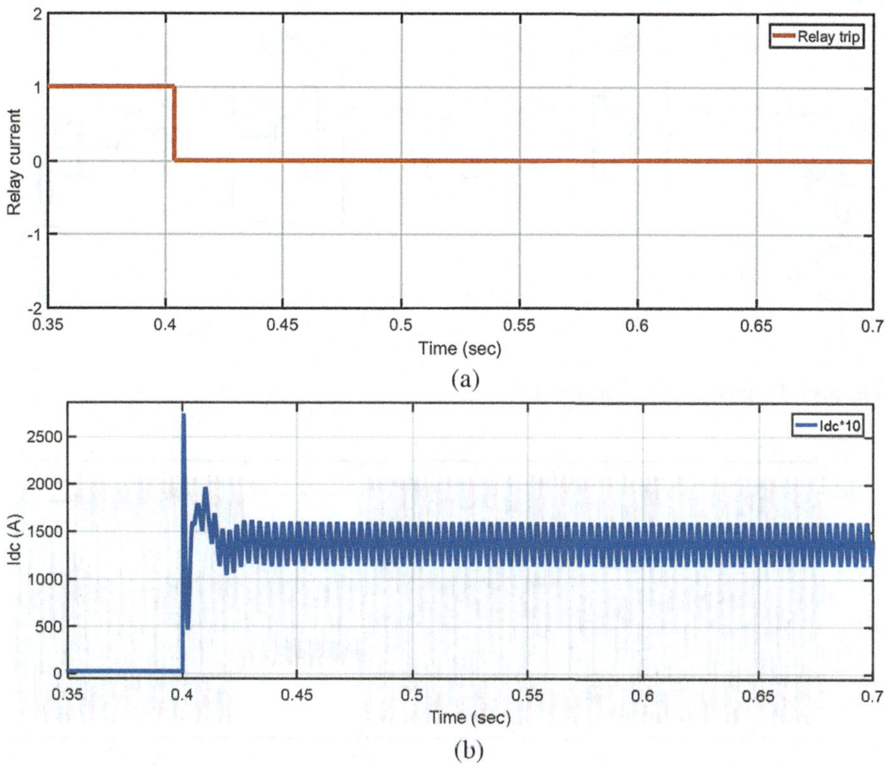

Fig. 5.18 Effect of circuit breaker operation on relay. (a) Response of relay after connected to a circuit breaker. (b) DC fault current

5.8 Transient Grid Faults: The Voltage Sags (Dips)

A voltage sag, also called a voltage dip, is a sudden reduction of the grid voltage at the PCC, generally between 10% and 90% of the rated value, during a period lasting from half a cycle to a few seconds. The sag faults are applied at grid A at time 0.4 to 0.5 s and clear fault at 0.5 s. It is possible to distinguish between different types of voltage sags as shown in Fig. 5.19.

5.8.1 Sag Type A: Three Phases with Ground Fault

Figure 5.20 shows the sag type A; this fault occurs at grid 1 in time from 0.4 to 0.5 s. When the fault occurs, the voltage drops from normal voltage to 120 V. After clean fault, the system restarts fast due to the control system.

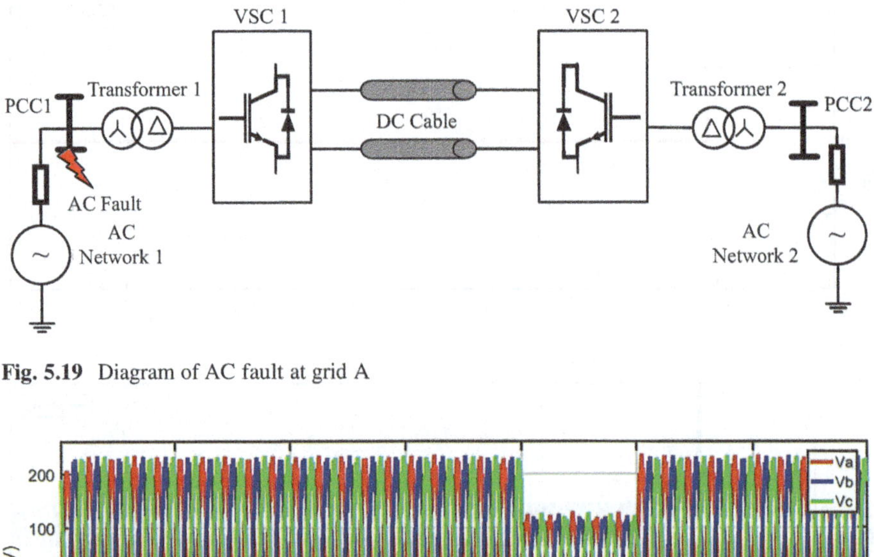

Fig. 5.19 Diagram of AC fault at grid A

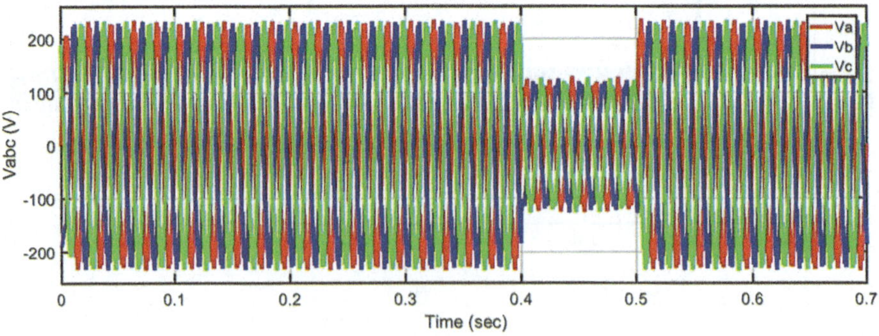

Fig. 5.20 Sag type A: three phases with ground fault

5.8.2 Sag Type B: Single-Phase-to-Ground Fault

Figure 5.21 shows the sag type B; this fault occurs at grid 1 in time from 0.4 to 0.5 s. When the fault occurs, the voltage at phase a drops from normal voltage to 120 v, but two other phases have constant phase jump angle. After clean fault, the system restarts fast due to the control system.

5.8.3 Sag Type C: Phase-to-Phase Fault

Figure 5.22 shows the sag type C; this fault occurs at grid 1 in time from 0.4 to 0.5 s. When the fault occurs, the voltage at phase a and phase b drop from normal voltage to 120 v, but other phase has different phase jump angle. After clean fault, the system restarts fast due to the control system.

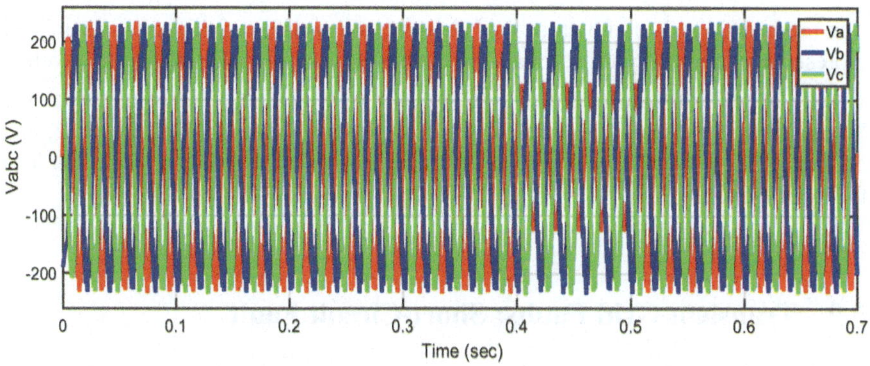

Fig. 5.21 Sag type B: single-phase-to-ground fault

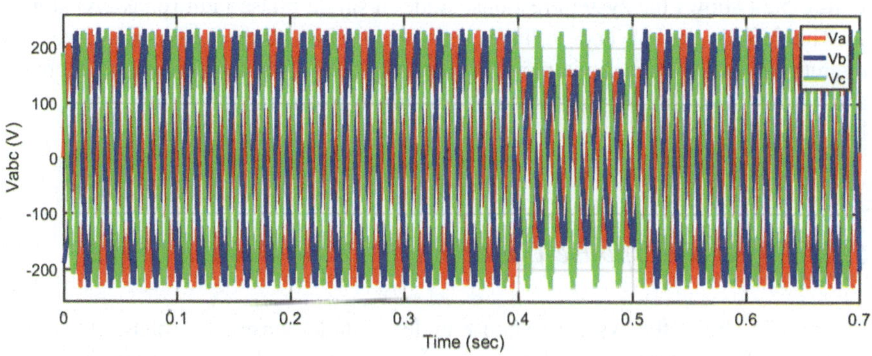

Fig. 5.22 Sag type C: phase-to-phase fault

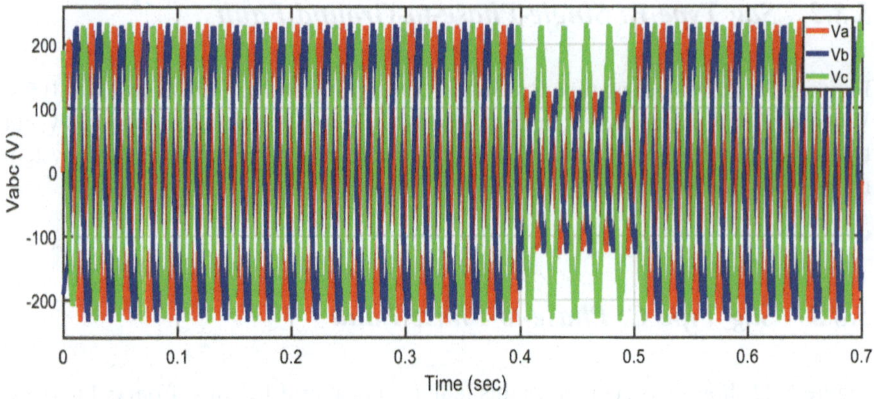

Fig. 5.23 Sag type D: two-phase-to-ground fault with $D = 0.5\angle 0^{\circ}$

5.8.4 Sag Type D: Two-Phase-to-Ground Fault

The product $\vec{D}\overrightarrow{Vsa^{+}}$ is known as the "characteristic voltage of the voltage sag" and represents either the phase voltage in phase-to-ground faults or the line-to-line voltage in phase-to-phase faults. Likewise, the phase angle of $\vec{D}$ is known as the "characteristic phase angle jump" of the voltage sag. Figures 5.20, 5.21, 5.22, and 5.23 shows four types of voltage sag. The voltage phasors during the grid fault lose the symmetry.

5.9 Transient Grid Faults: Short-Circuit Fault

5.9.1 System Response Under Single-Phase Fault at Grid 1

Figure 5.24 shows the system response under a single-phase fault (phase A) at grid 1. The VSC1 current and voltage waveforms show that the system faces an overcurrent and voltage dip in phase A (faulted phase). This fault created an increase and harmonics in the DC-link voltage due to an unbalance in the voltage and current like three-phase fault.

5.9.2 System Response Under Single-Phase Fault at Grid 2

Figure 5.25 shows the system response under a single-phase fault (phase A) at grid 2. The VSC2 current and voltage waveforms indicate that the system faces an unbalanced overcurrent and voltage dip in phase A (faulted phase). This fault

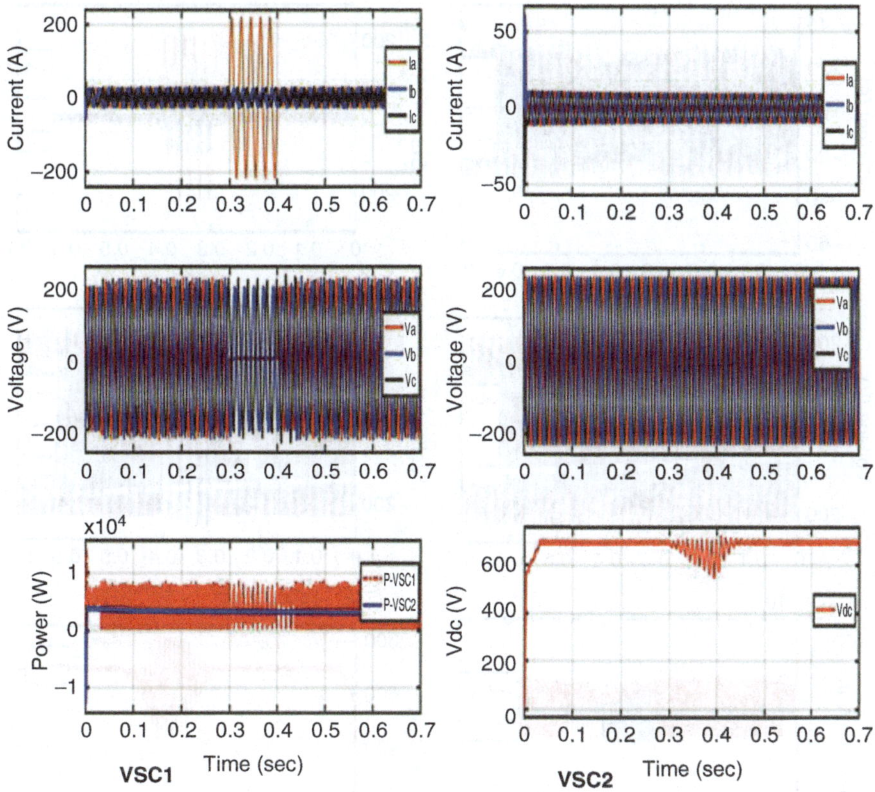

Fig. 5.24 System response under fault at grid 1; (**a**) current and voltage at *PCC* point of grid 1, (**b**) current and voltage at PCC point of grid2, (**c**) sending and receiving active power, and (**d**) DC voltage link

created a decrease and rich of harmonics in the DC-link voltage due to an unbalance in the voltage and current that is reflected on the DC link, similarly for the previous fault scenario.

5.9.3 System Response Under Three-Phase Fault at Grid 1

Figure 5.26 shows the performance of HVDC under a three-phase fault at grid 1. A three-phase fault is applied at time $t = 0.3$, for 1 s fault duration. The VSC1 current and voltage waveforms show that the system faces an overcurrent and voltage dip; this fault has led to increasing the DC-link voltage as the power transmitted from VSC2 is accumulated at the DC link due to the fault at grid 1 (which made the power at the receiving end almost zero). After the fault had been cleared, the controller

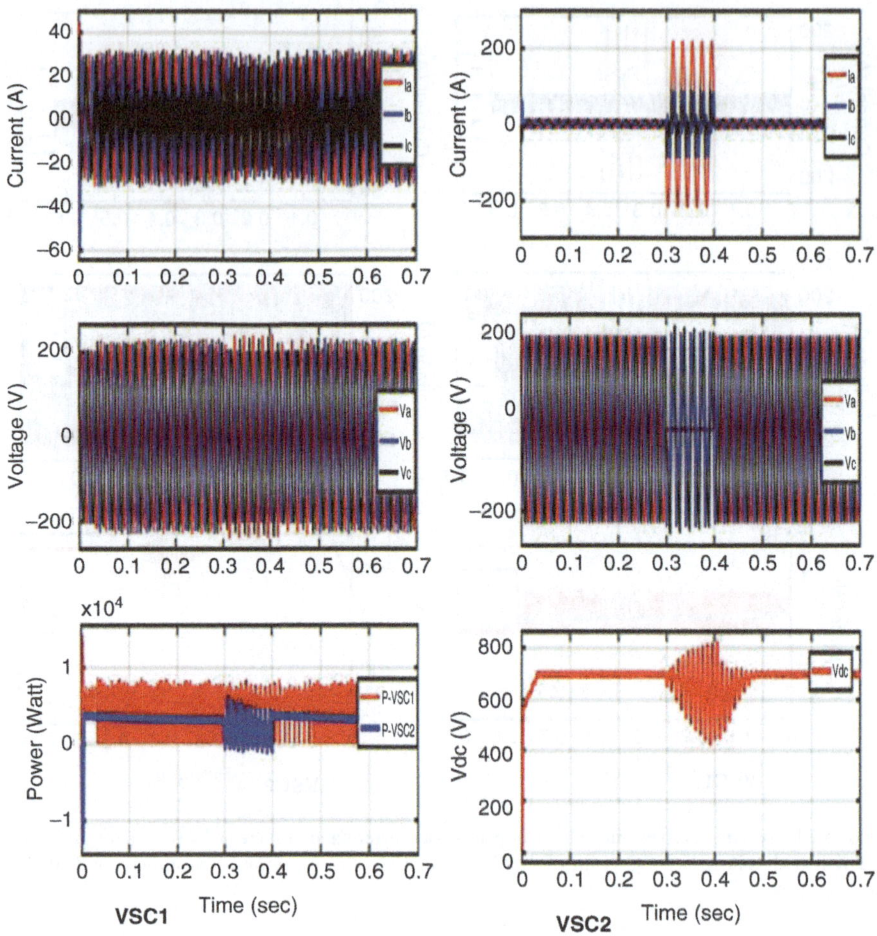

Fig. 5.25 System response under fault at grid 2; (**a**) current and voltage at *PCC* point of grid 1, (**b**) current and voltage at PCC point of grid2, (**c**) sending and receiving active power, and (**d**) DC voltage link

brought the system to the steady-state condition. However, there is no impact of the fault on the other end (grid 2) since DC link acts as a buffer to isolate both ends.

5.9.4 System Response Under Three-Phase Fault at Grid 2

Figure 5.27 shows the system response under a three-phase fault at grid 2. The VSC2 current and voltage waveforms show that the system faces an overcurrent and voltage dip. This fault has led to decreasing the DC-link voltage because the power fed to the DC is reduced due to the fault. This decrease is manifested as a

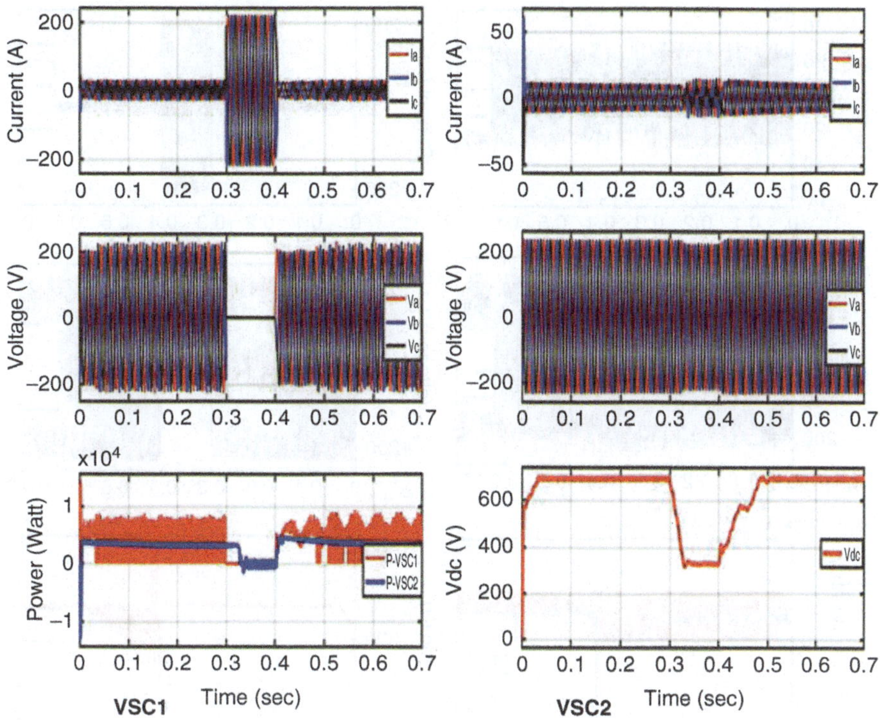

Fig. 5.26 System response under fault at grid 1; (a) current and voltage at *PCC* point of grid 1, (b) current and voltage at PCC point of grid2, (c) sending and receiving active power, and (d) DC voltage link

decrease in DC link and in the power (current) at the receiving end. In comparison, fault at grid 2, where the active power is controlled, only reduces the current. However, voltages are still not influenced in both cases.

5.10 Control Strategies of Transient Grid Faults: Short-Circuit Single-Line-to-Ground Fault (SL-G)

5.10.1 BPSC (Balanced Positive Sequence Control)

In the presence of the same default, BPSC allows to have sinusoidal currents, balanced and in-phase with the corresponding voltages. The current amplitude is increased to maintain the average active power injected to the PCC during the voltage dip. Only, important second harmonic oscillations were superimposed upon both active and reactive powers, which is detrimental to stability of the power system Fig. 5.28.

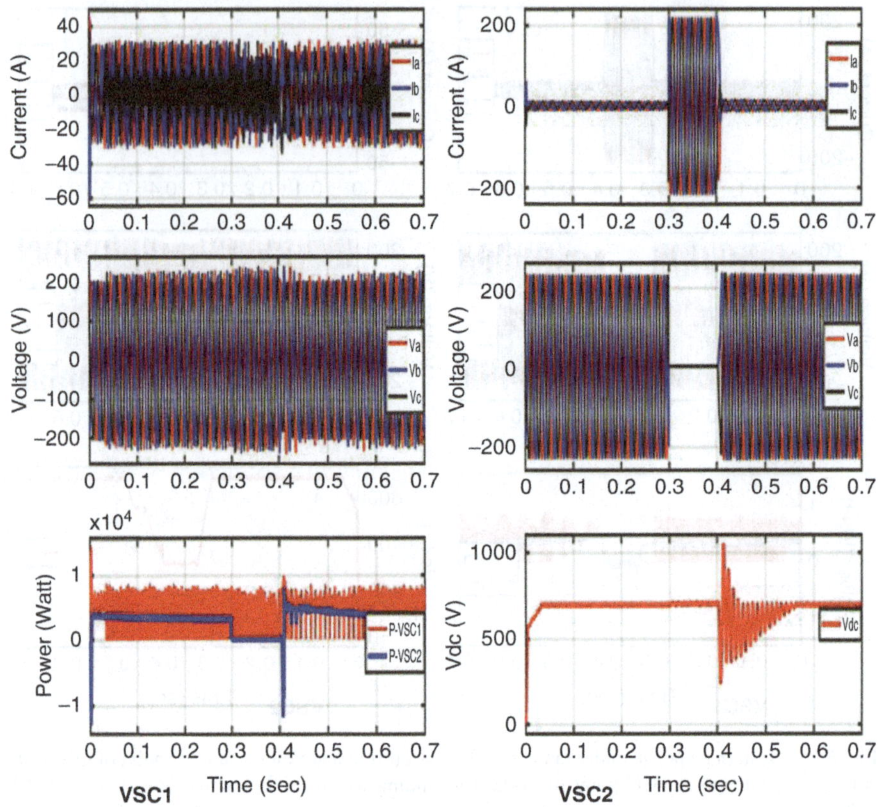

Fig. 5.27 System response under fault at grid 2; (**a**) current and voltage at *PCC* point of grid 1, (**b**) current and voltage at PCC point of grid2, (**c**) sending and receiving active power, and (**d**) DC voltage link

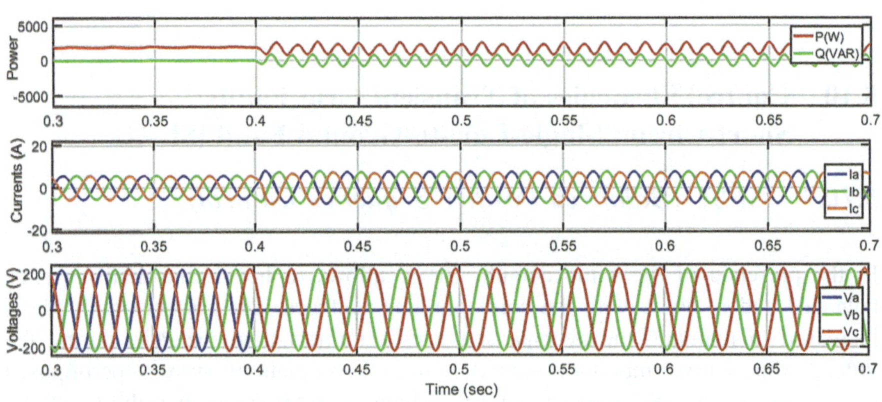

Fig. 5.28 Grid voltages, currents, active and reactive powers using BPSC control

5.10.2 IARC (Instantaneous Active-Reactive Control)

The application of the IARC control technique shows that although the active and reactive powers are kept constant even in case of fault, the currents injected to the grid are unbalanced and very harmonic polluted currents as shown in Fig. 5.29.

5.10.3 PNSC (Positive Negative Sequence Control)

The application of the PNSC technique shows that although the reactive powers are kept constant but the active power are oscillations in case of fault, currents injected to the grid are balanced and low harmonic polluted currents as shown in Fig. 5.30.

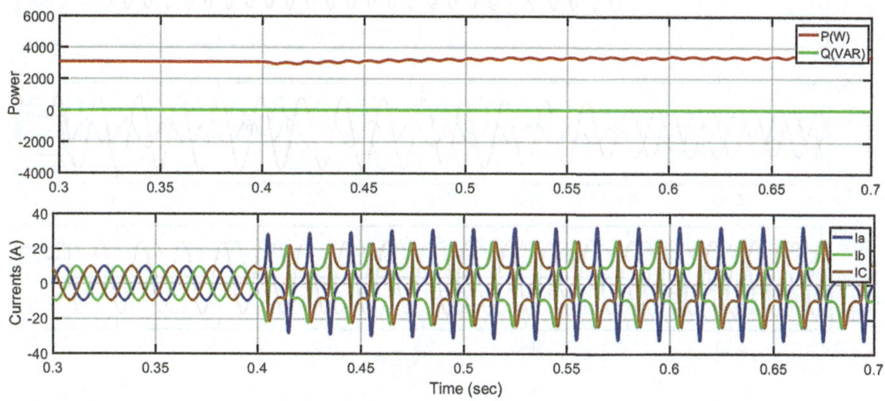

Fig. 5.29 Grid currents, active and reactive powers using IARC control

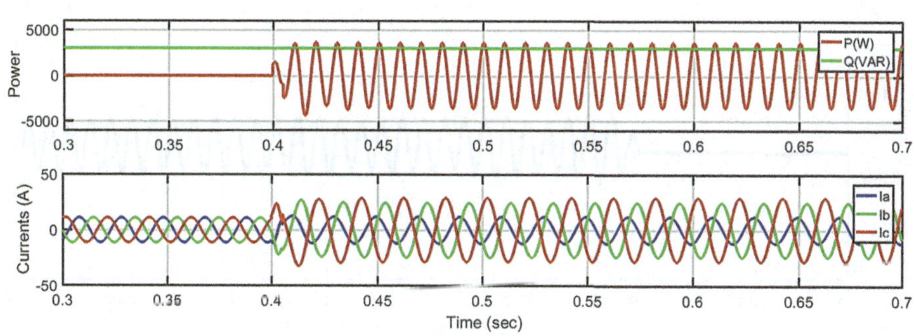

Fig. 5.30 Grid currents, active and reactive powers using PNSC control

5.10.4 AARC (Average Active-Reactive Control)

For the AARC method, the active power reference is maintained equal to its reference $P^{*}g$ and that of the reactive power is kept at zero. It can be proven that the injected currents are assumed sinusoidal unbalanced in amplitude when unbalance occurs (Fig. 5.31). These currents are proportional to the point of common coupling voltages. However, the active power reveals significant oscillations at twice the fundamental frequency. Moreover, this technique is able to eliminate the reactive power oscillations.

The results of the simulation are shown in Fig. 5.32. It can be observed that the oscillations of DC side voltage are successfully eliminated with the controller. Similar results are obtained for unbalanced faults applied at VSC1 station.

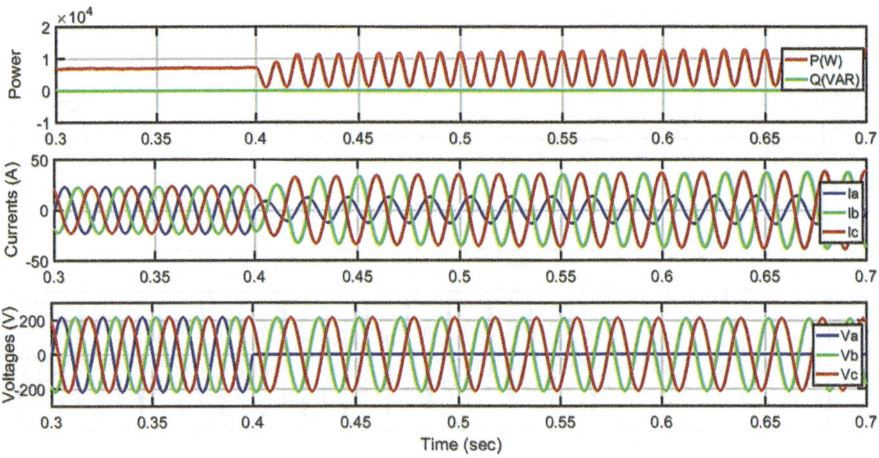

Fig. 5.31 Grid currents, active and reactive powers using AARC control

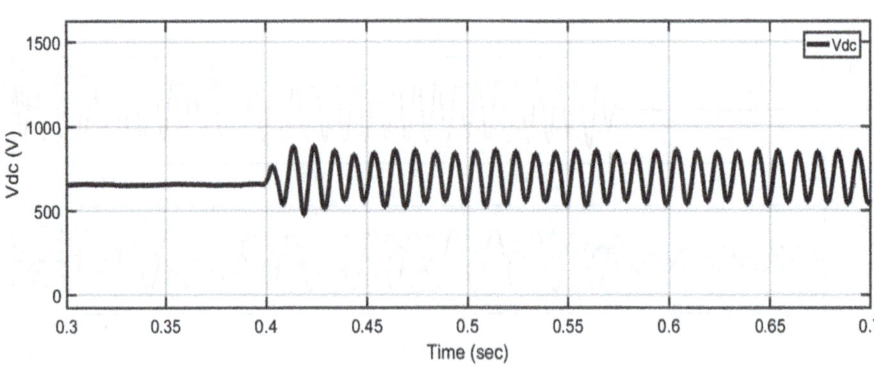

Fig. 5.32 DC side voltage and for reference distribution strategies during unbalance fault

5.11 Control Strategies of Transient Grid Faults: Voltage Sag Phase-to-Phase Fault (L-L)

5.11.1 *BPSC (Balanced Positive Sequence Control)*

In the presence of the same default, BPSC allows to have sinusoidal currents, balanced and in-phase with the corresponding voltages. The current amplitude is increased to maintain the average active power injected to the PCC during the voltage dip-type phase-to-phase sage fault. Only, important second harmonic oscillations was superimposed upon both active and reactive powers, which is detrimental to stability of the power system as shown in Fig. 5.33.

5.11.2 *PNSC (Positive Negative Sequence Control)*

The application of the PNSC technique shows that although the reactive powers are kept constant but the active power are oscillations in case of fault, currents injected to the grid are balanced and low harmonic polluted currents as shown in Fig. 5.34.

5.11.3 *AARC (Average Active-Reactive Control)*

For the AARC method, the active power reference is maintained equal to its reference $P^{*}g$ and that of the reactive power is kept at zero. It can be proven that the injected currents are assumed sinusoidal unbalanced in amplitude when

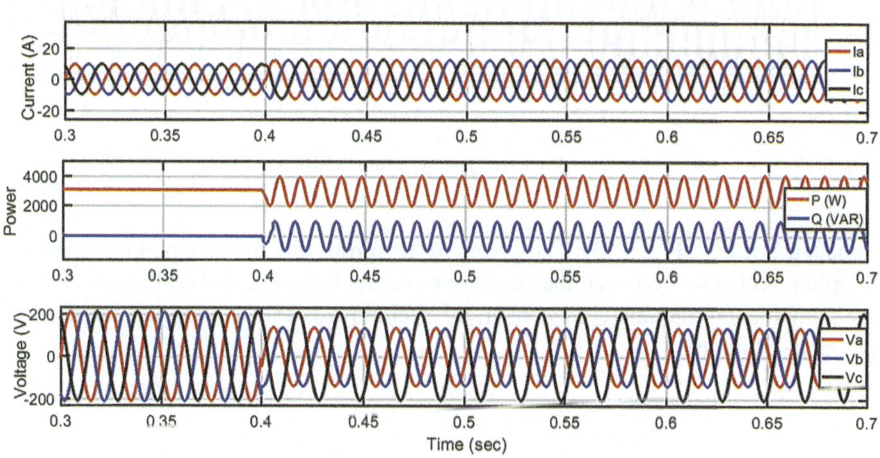

Fig. 5.33 Grid voltages, currents, active and reactive powers using BPSC control

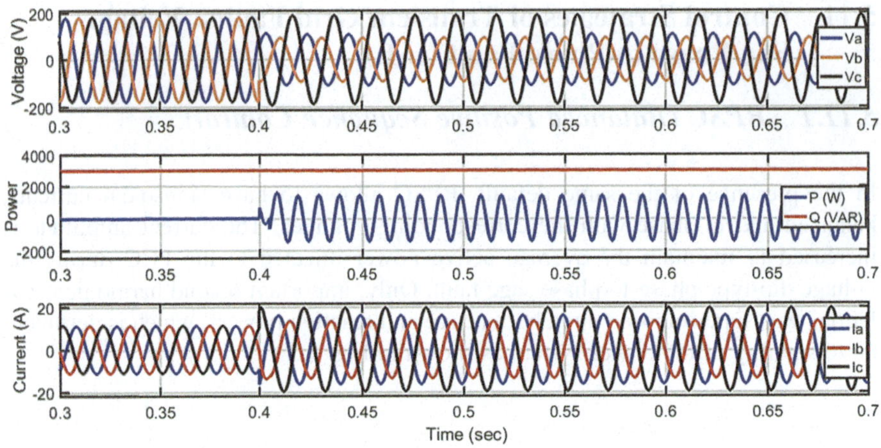

Fig. 5.34 Grid currents, active and reactive powers using PNSC control

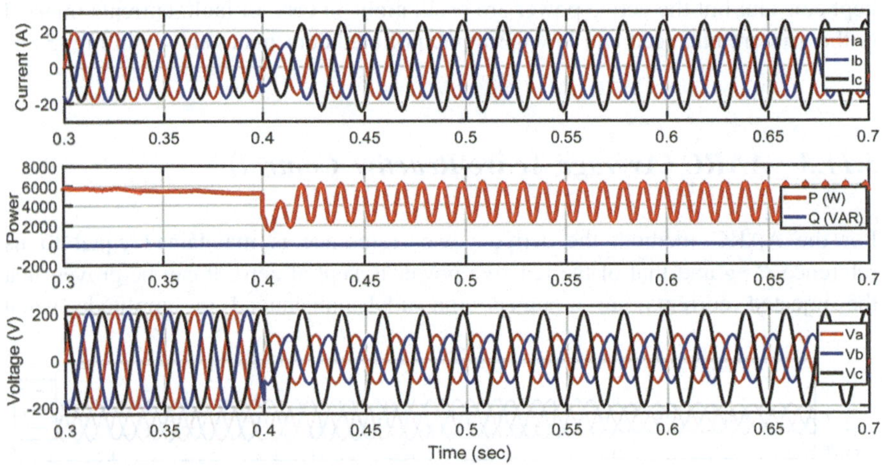

Fig. 5.35 Grid currents, active and reactive powers using AARC control

unbalance occurs (Fig. 5.35). These currents are proportional to the point of common coupling voltages. However, the active power reveals significant oscillations at twice the fundamental frequency. Moreover, this technique is able to eliminate the reactive power oscillations.

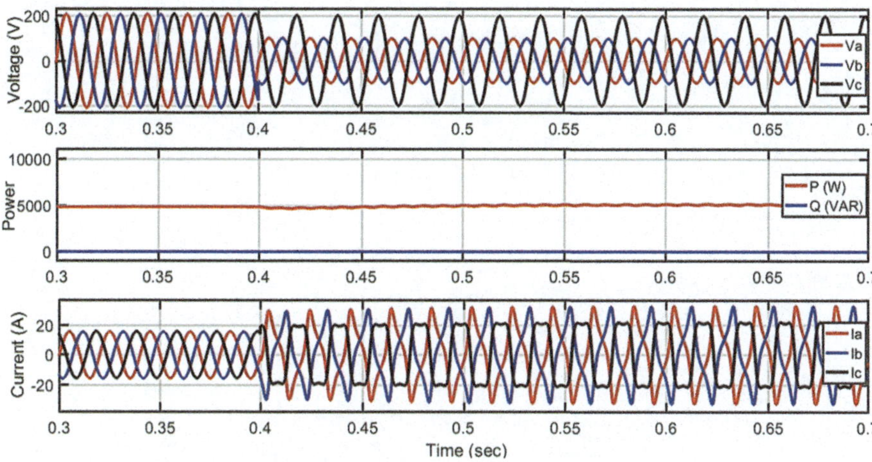

Fig. 5.36 Grid currents, active and reactive powers using IARC control

5.11.4 IARC (Instantaneous Active-Reactive Control)

The application of the IARC control technique shows that although the active and reactive powers are kept constant even in case of fault, the currents injected to the grid are unbalanced and very harmonic polluted currents as shown in Fig. 5.36.

Chapter 6
Experimental Investigation for HVDC System

6.1 Introduction

In recent years, the wide array of digital devices is available for developing drive system. Unlike the analog controllers, the digital controllers allow implementation of sophisticated operating algorithms, offering significant technical advantages, thanks to their ability of fast processing of vast data, high operational flexibility, and ease of integration into automated industrial systems. In recent years, the digital signal processors (DSPs1104) have been enjoying increased interest and implementation in control of electrical systems. The advantages of the DSP1104 are flexibility, insensitivity to aging effects and for thermal drifts, ease of implementation and upgrade, and, on the other hand, it is available at low prices. As an integrated part of this work, an attempt is made to verify the control scheme experimentally. This chapter describes the details of all the hardware that are used to verify the effectiveness of the proposed scheme.

6.2 Description of the Experimental Setup

A series of experimental investigations have been carried out to verify the validity of the control scheme of inverter system. The results were obtained at different operating points including transient steady state. All the results during the tests were displayed through the A/D converters on the DSP1104 controller board and recorded using the trace of the digital signal processor. The major parts of the experimental arrangement are illustrated in Figs. 6.1 and 6.2. Some circuits such as protection, driving, and measuring circuits are implemented to complete the operation of the control system. The following sections describe the main components of the system in detail.

N. F. Ibrahim, S. S. Dessouky, *Design and Implementation of Voltage Source Converters in HVDC Systems*, Power Systems,
https://doi.org/10.1007/978-3-030-51661-1_6

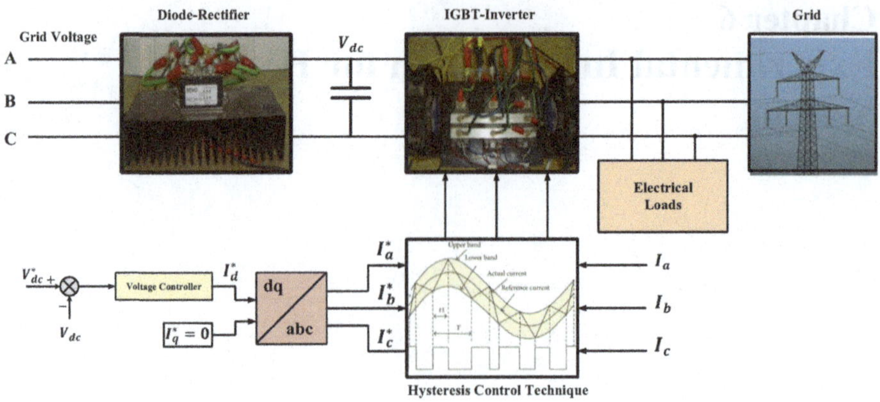

Fig. 6.1 Experimental test

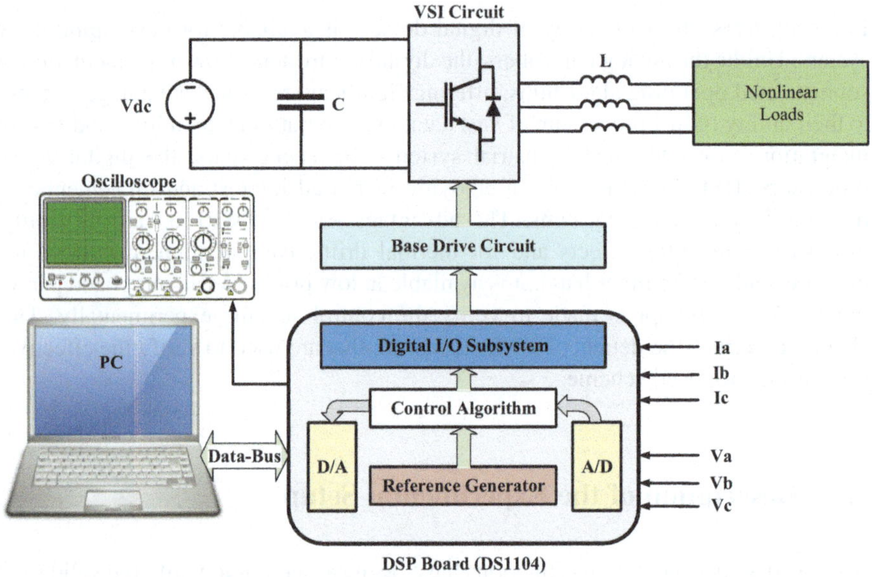

Fig. 6.2 Digital signal processing (DSP) system with experimental test

6.2.1 Main Power Circuit

As shown in Fig. 6.3, the power circuit is constructed using three-phase uncontrolled rectifier with variable resistance at its DC side which is used as a nonlinear load connected with the three-phase supply. A three-phase autotransformer has been used to supply the system with a variable AC voltage. The VSI is connected in parallel with the nonlinear load through capacitor C (capacitor bank) that carries the input ripple current of the inverter and the main reactive energy storage element. The DC

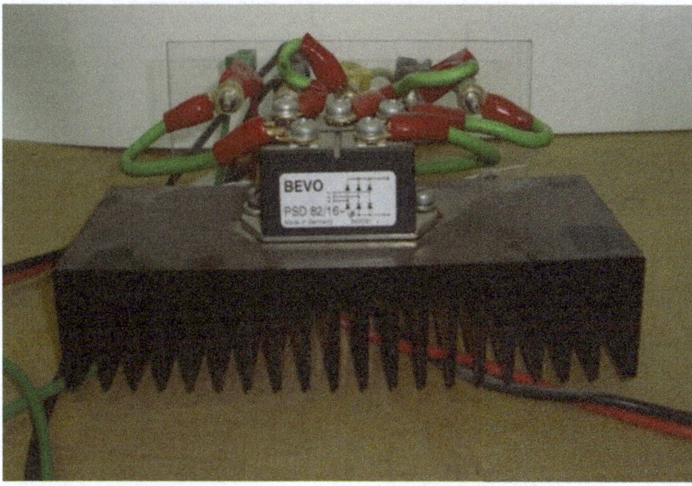

Fig. 6.3 Real view of the three-phase diode bridge rectifier

capacitor provides a constant DC voltage and the real power necessary to cover the losses of the system. The inductors L1, L2, and L3 perform the voltage boost operation in combination with the capacitor. Figure 6.3 shows the real view of the three-phase rectifier.

6.2.2 IGBT Three-Phase Voltage Source Inverter (VSI) Circuit

Voltage source inverter (VSI) contains three-phase bridge isolated gate bipolar transistor (IGBT) [type: CM50DY-24H] antiparalleling the diodes. The data sheet of the IGBT module is provided in Appendix A. Voltage source inverter (VSI) was used to convert the DC bus voltage into three-phase AC voltages with variable frequency. The VSI power circuit is shown in Fig. 6.4. The IGBT-VSI is controlled by PWM signals generated by the controller board. The real views of the three-phase IGBT power inverter, the three-phase reactor (L1, L2, L3), and the capacitor bank are shown in Figs. 6.5, 6.6, and 6.7, respectively.

6.2.3 Snubber Circuit

This circuit is used in conjunction with IGBT to protect the semiconductor devices, and this function is performed by:

- Limiting device voltages during turn-off transients
- Limiting device currents during turn-on transients

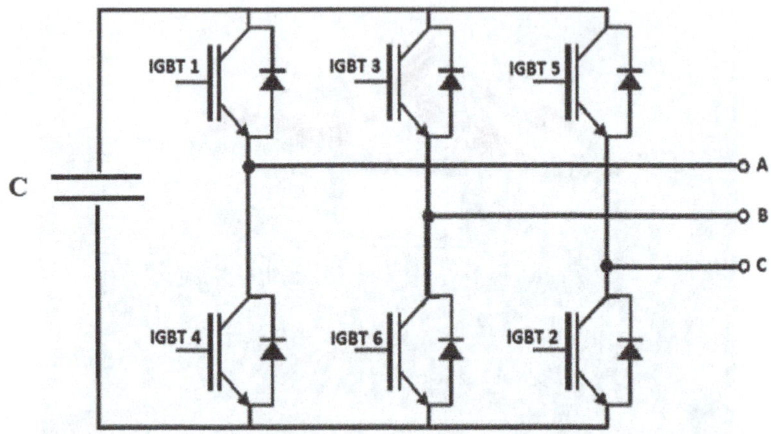

Fig. 6.4 IGBT power inverter switches

Fig. 6.5 Real view of the three-phase IGBT power inverter

Fig. 6.6 Real view of the three-phase reactor

Fig. 6.7 Real view of the capacitor bank

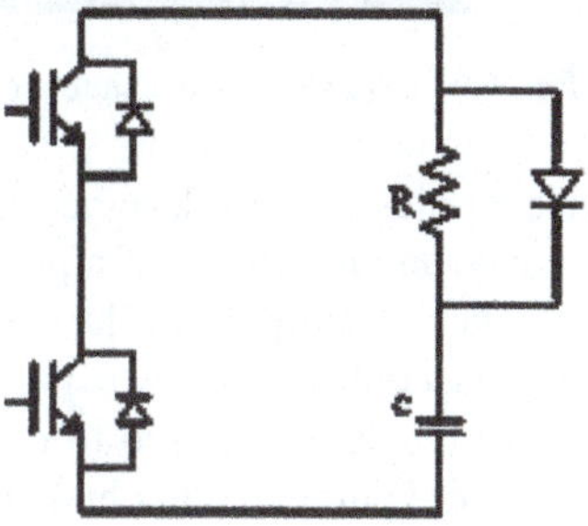

Fig. 6.8 Snubber circuit (one inverter leg)

- Limiting the rate of rise (di/dt) of currents through the semiconductor device at device turn-on
- Limiting the rate of rise (dv/dt) of voltages through the semiconductor device at device turn-off

The schematic diagram and the real view of the common snubber circuit are shown in Figs. 6.8 and 6.9, respectively.

In the proposed control system, the snubber circuit has been used to achieve the following advantages:

- It is used with inverter with carrier frequency of up to 2 KHZ.
- Reduce the switching loss.
- Suppress the surge voltages.

6.2.4 Inverter Driving Circuit

The gate drive circuitry of a power inverter forms an important interface between the high-power electronic and the digital or intelligent control processing stages. It is therefore of utmost importance that this interface between the control and power

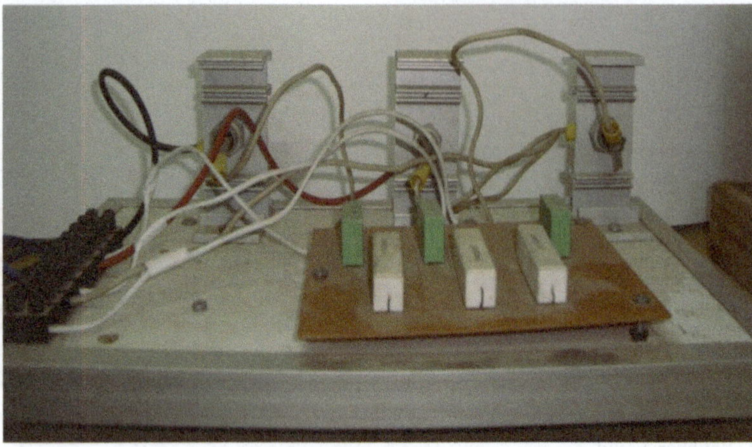

Fig. 6.9 Real view of the snubber circuit

electronics is well designed, since it can have a substantial impact on the performance and reliability of a power electronic system.

The first purpose of the driving circuit is to give the isolation between the power and the control sides by using opto-coupler. The second purpose is to connect the six base signals required to turn on or to turn off the power transistors of the inverter. The third purpose is to obtain a suitable dead time between each two transistors in the same inverter arm in order to prevent short circuit on the DC supply when the two transistors on the same arm are conducted at the same time. Thus, a base drive circuit is essential for the inverter to provide an isolation and appropriate voltage to the six IGBTs of the inverter.

The real view of the base drive circuit is shown in Fig. 6.10. The details of this circuit are illustrated in Appendix B. The chip SN7407N has been used as a level shifter that shifts the voltage level from +5 V to +15 V. The chip HP2531 is an opto-coupler that has been used to provide isolation between the logic circuit and the power circuit of the inverter. Thus, a base drive circuit is essential for the inverter to provide an isolation and appropriate voltage to the six IGBTs of the inverter.

The chip IR2130 is the main drive that provides six driving pulses for the six switches of the inverter. In order to provide +20 V isolated power to the opto-coupler and the drive, an isolated power supply has been built which is also shown in Appendix C.

6.2.5 *Current Measurement*

In order to implement feedback current control, the three-phase output currents of the inverter should be measured online and the measured values supplied to the controller via its ADC converters.

Fig. 6.10 Real view of the base drive circuit

Three-phase load currents to be used in the extraction reference current must also be measured online. In this work, we assume that the three phases are balanced; thus only two current phases (I_a and I_b) are required to be measured, and the third one can be calculated by:

$$I_c = -(I_a + I_b) \tag{6.1}$$

The two-phase currents were measured using two current transducers (LEM Module LA 25-NP). These transducers use a Hall Effect Principle, providing electrical isolation between the power circuit and the controller. Each transducer was designed to measure a maximum of 25 A with a sensitivity of 100 mV/A. The output current signal of these sensors is converted to voltage by connecting a resistor between output of the sensors and ground. Output voltage can be scaled by selecting various resistor values and fed to the DSP board through the A/D converter. The per phase circuit is shown in Fig. 6.11 and data sheets are given in Appendix C.

6.2.6 *Voltage Measurement*

The voltage can be measured using LEM Module LV 25-P voltage transducers. These transducers also function according to the Hall Effect Principle, so they provide electrical isolation between the power circuit and the controller. Figure 6.12 presents block diagram of the voltage transducer, while Fig. 6.13 shows a real view of both current and voltage transducers (3-LA 25-NP and 3-LV 25-P). The measuring range can be changed by varying the measuring resistance.

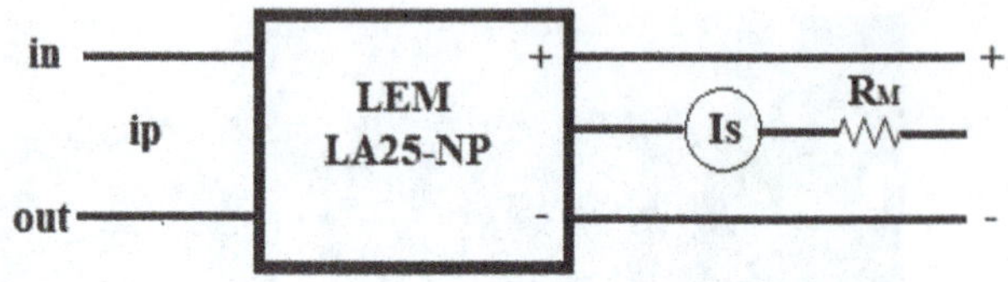

Fig. 6.11 Current transducer

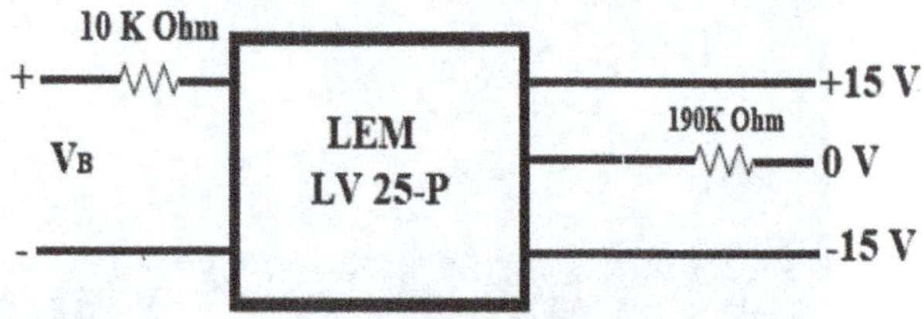

Fig. 6.12 Voltage transducer

Fig. 6.13 Real view of current and voltage transducers

6.2.7 *Auxiliary Power Supply Circuits*

There are two power supply circuits:

- Common power supply circuit which gives +15 and −15 V DC to the hall sensor circuits
- Base drive circuit power supply, which gives three voltage levels, +5, +15, and 20 V DC, as required for the circuit indicated in its data sheet

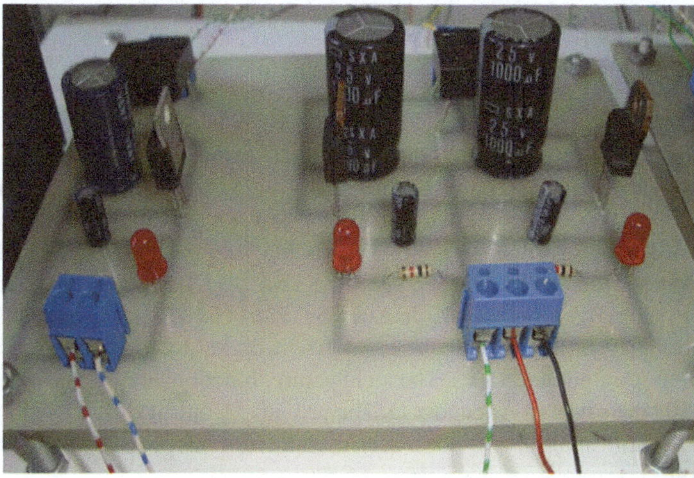

Fig. 6.14 Experimental setup of the common power supply circuit

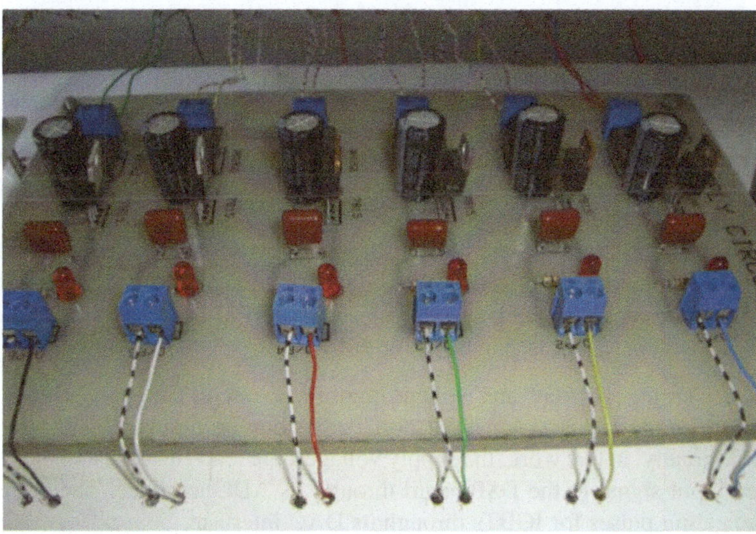

Fig. 6.15 Experimental setup of the base drive power supply circuit

The real view of both the common power supply and base drive power supply circuits are shown in Figs. 6.14 and 6.15, respectively.

6.2.8 DSP1104 Controller Board

The inverter system was controlled using DSPACEDS1104 controller board, which is installed on a personal computer. This board is specifically designed for the development of high-speed multivariable digital controllers and real-time simulations. DSP1104 controller board is placed in the PC slot on the main board of PC. The DSP1104 contains a main processor and a slave digital signal processor (DSP1104). The main processor is a 603 power PC, running at 250MHZ with 32 MB of SDRAM, and the slave DSP1104 is a TMS320F240, Texas Instrument's floating point DSP, 20MHZ CPU clock. It contains four 16-bit analog-to-digital (A/D) channels, four 12-bit A/D channels, eight 16-bit digital-to-analog (D/A) channels, and other input/output interfaces. The architecture, functional units, and data sheet of the DSP1104 are given in Appendix D. The provided Control Desk Developer along with MATLAB/Simulink provides user-friendly interface to system control and observation.

There are three different ways to connect external devices to the DSP1104. To access the I/O units of the master PCC and the slave DSP, connect external devices:

- To the 100-pin I/O connector P1 of the DSP1104.
- To the two 50-pin Sub-D connectors P1A and P1B; those are included in the DSP1104 hardware package.
- To the optional connector panel CP1104 or the optional combined connector/LED panel CLP1104, which provides an additional array of LEDs indicating the states of the digital signals. Pin assignment of I/O connectors (P1 and Sub-D connectors) is given in Appendix E.
- To make the interconnection of both input signals (current/voltage transducers and incremental encoder) and output signals (inverter IGBT switches) with the DSP1104 board simple and easy, a connection port of DSP is extended through an interface board that is shown in Fig. 6.16.

We can arrange current and voltage transducer circuit, interface circuit (IR), power supply (5, 15, 20), and DSP port connection in a stand as illustrated in Fig. 6.17. Finally, in this work, the supply voltages, the load currents, and the current filter are input signal to the DSP board through its ADC interface. The outputs of DSP are gating pulses for IGBTs through its DAC interface.

6.3 Experimental Results and Discussion

Figures 6.18, 6.19, 6.20, 6.21, 6.22, 6.23, 6.24, 6.25, 6.26, 6.27, 6.28, 6.29, 6.30, and 6.31 depict the experimental test results of the prototype for proposed method of control to inverter and filter under a three-phase nonlinear load with a three-phase rectifier load at constant resistors. L parameters of the three-phase inductive are 12 mH per phase. R parameters of the three-phase rectifier load are 15 ohm. The

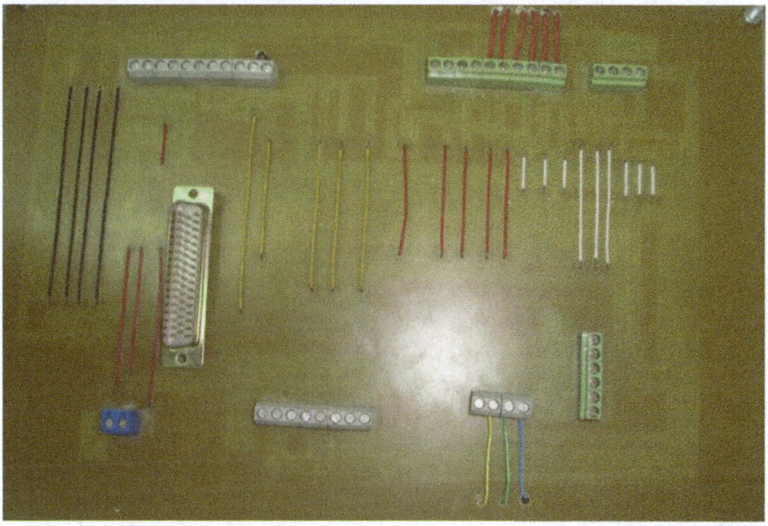

Fig. 6.16 DSP port connection

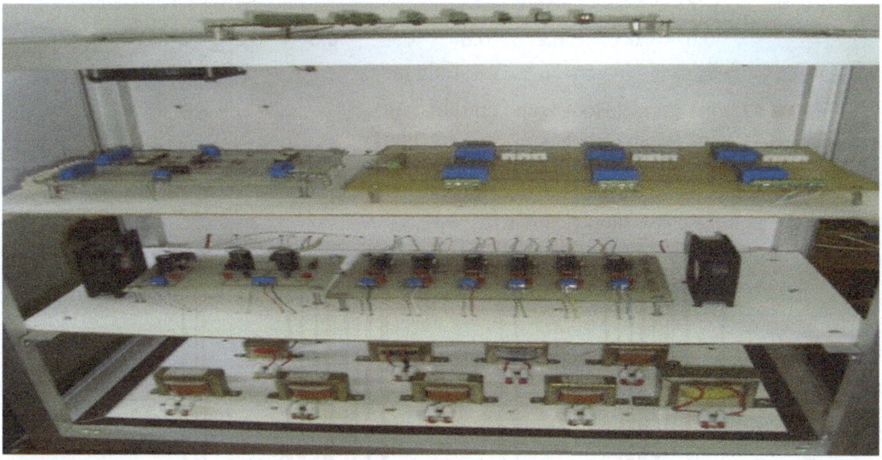

Fig. 6.17 A stand collecting measurement, DSP port connection, auxiliary power supply, and base drive circuit boards

various waveforms show the three-phase supply voltage, the three-phase supply current, the three-phase current before and after filter, the supply current for one phase superimposed with respective phase voltage before and after filter, the gate pulses for inverter switches, the current control response under change, and the three-phase load currents and voltages and DC-link voltage. Through experimental results, it is revealed that the supply currents after filter were found to be sinusoidal and in phase with the supply voltage, which was consistent with the simulation

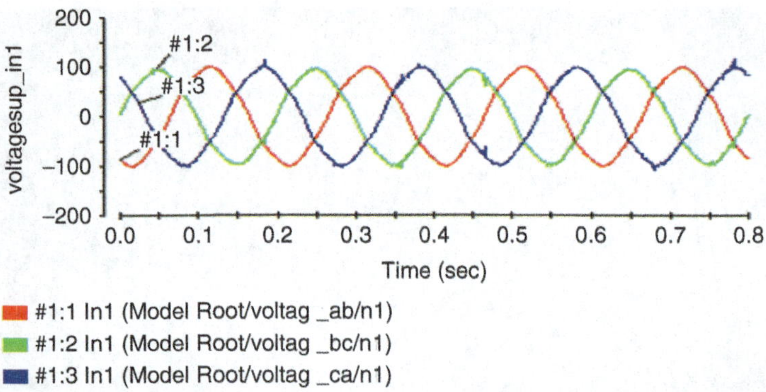

Fig. 6.18 The three-phase supply voltage

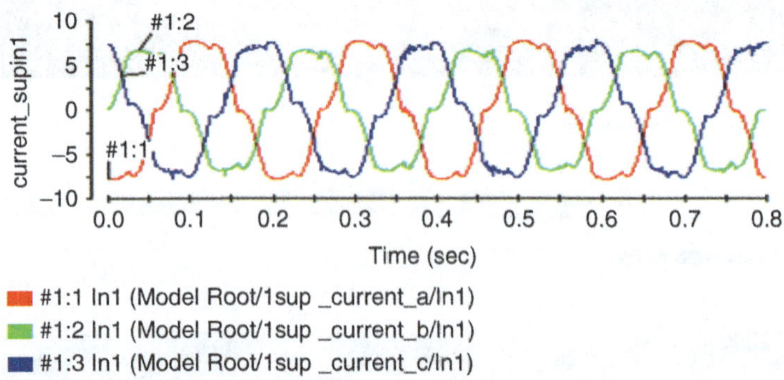

Fig. 6.19 The three-phase supply current

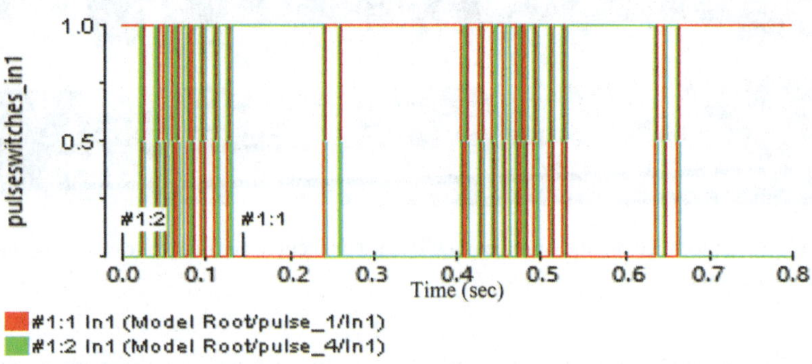

Fig. 6.20 Gate pulses for inverter switches

shown in the previous chapter. For this test, the outcome indicates that the proposed method is useful for the control studies under different operating conditions, and it indicates the effectiveness of the proposed controller for eliminating the harmonics.

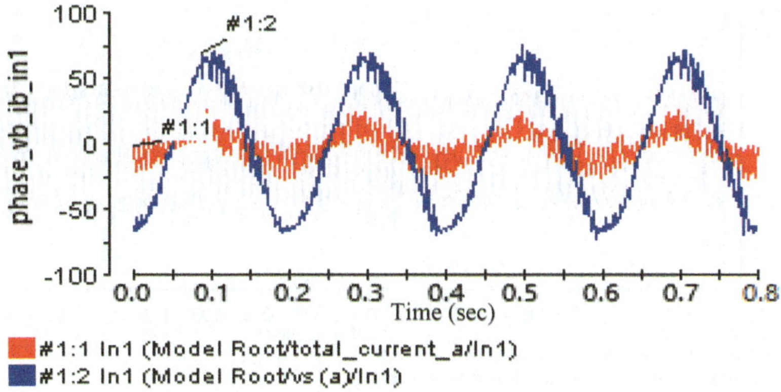

Fig. 6.21 Supply current and supply voltage

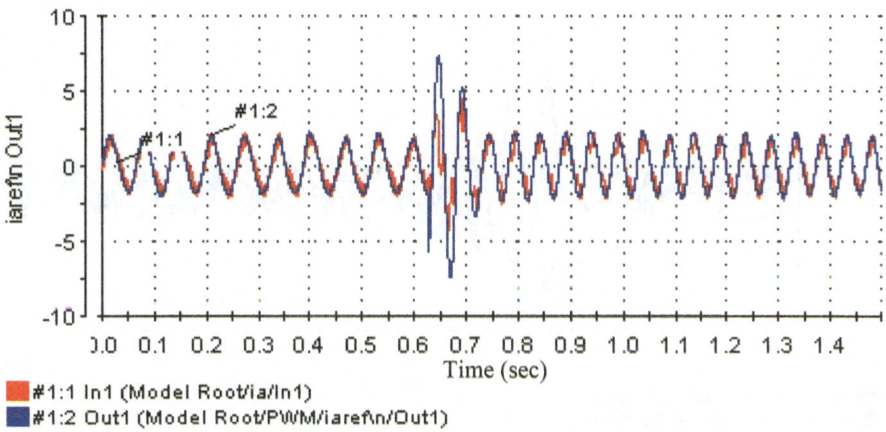

Fig. 6.22 Current control responses under change

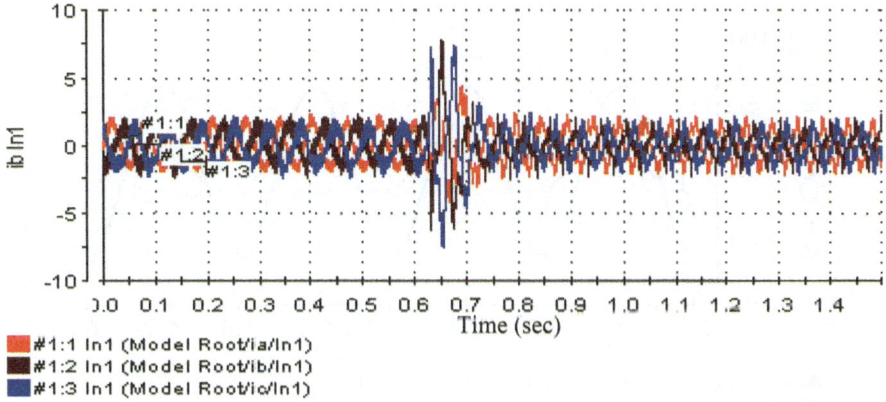

Fig. 6.23 The three-phase load current

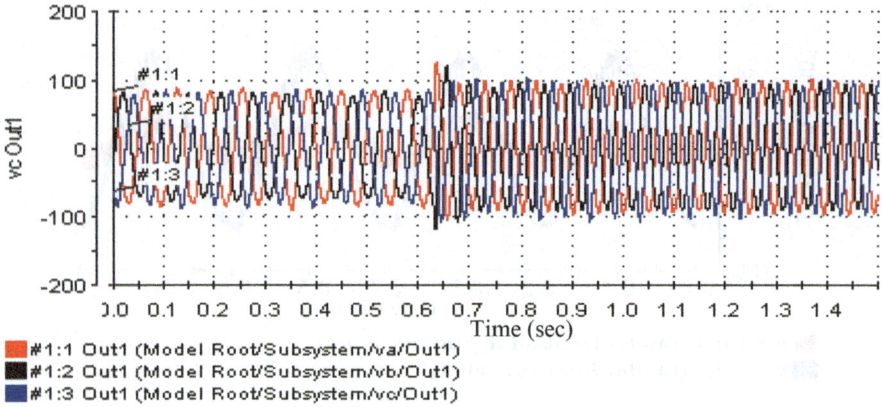

Fig. 6.24 The three-phase output voltage to inverter

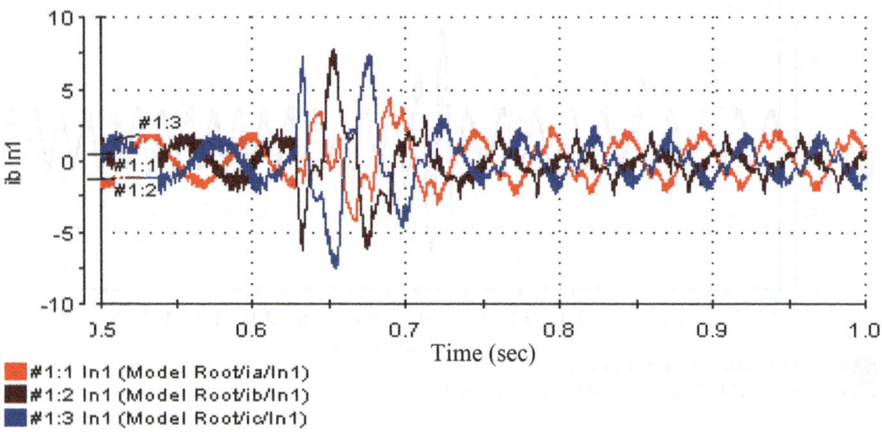

Fig. 6.25 The three-phase load current

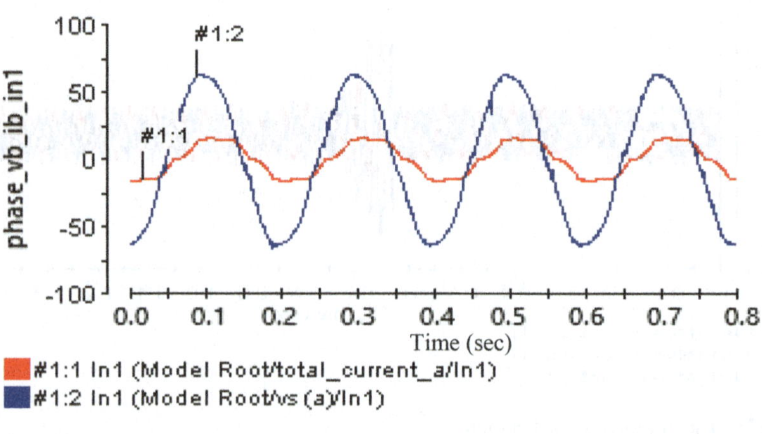

Fig. 6.26 Supply current and supply voltage before filter

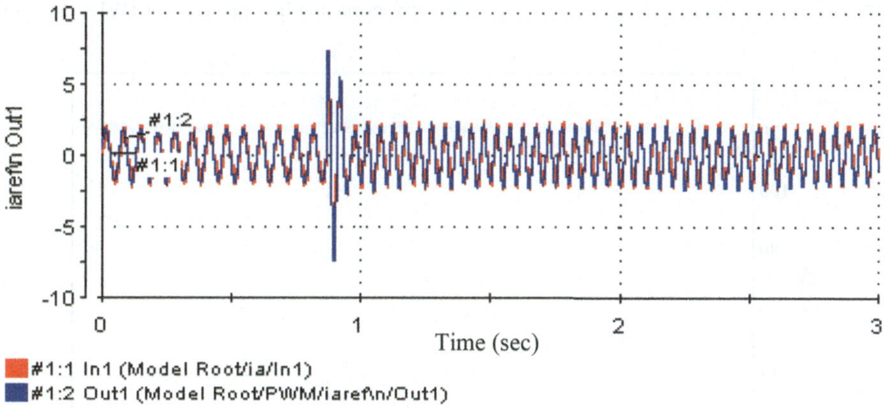

Fig. 6.27 Current control responses under change

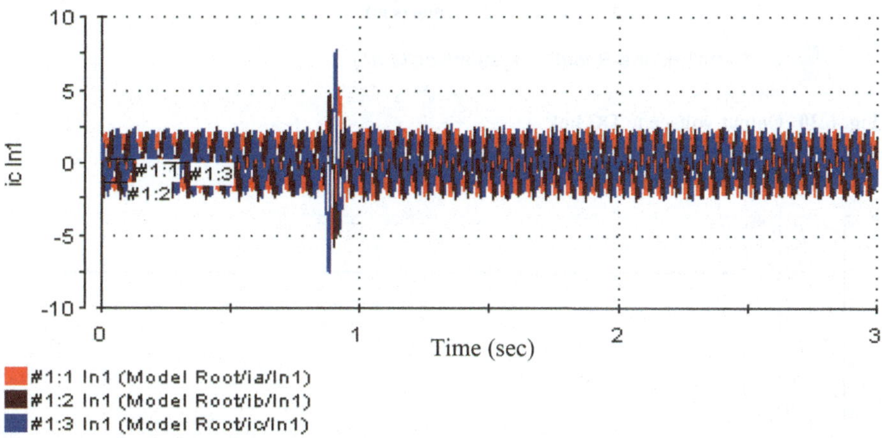

Fig. 6.28 The three-phase load current

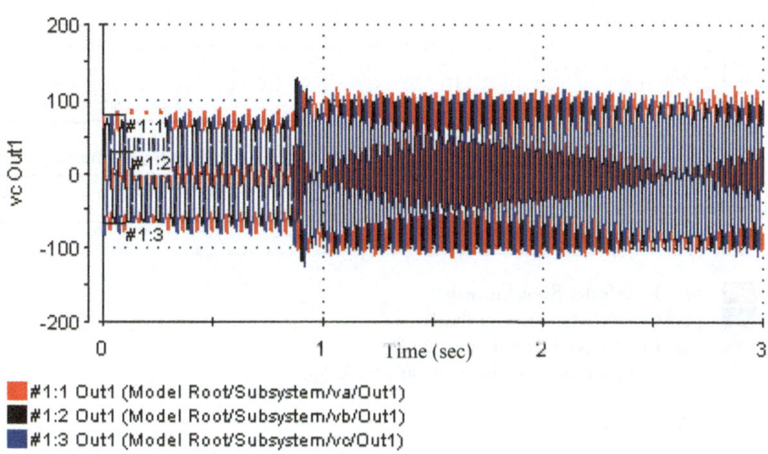

Fig. 6.29 The three-phase output voltage to inverter

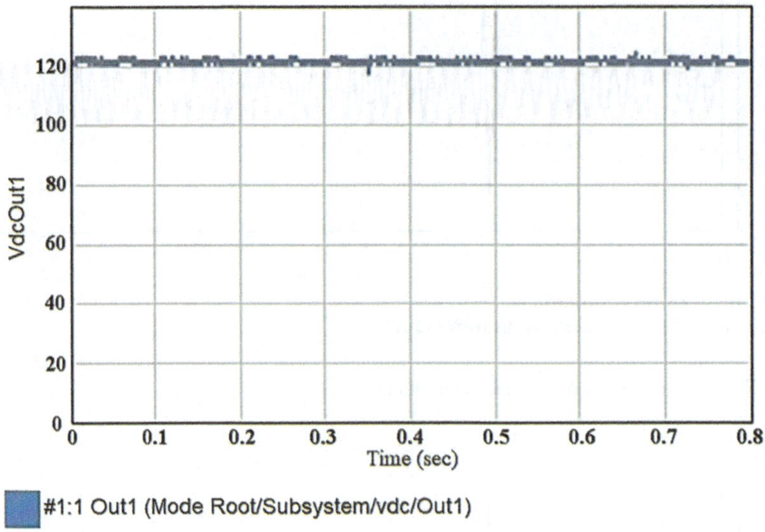

Fig. 6.30 Output voltage to DC link

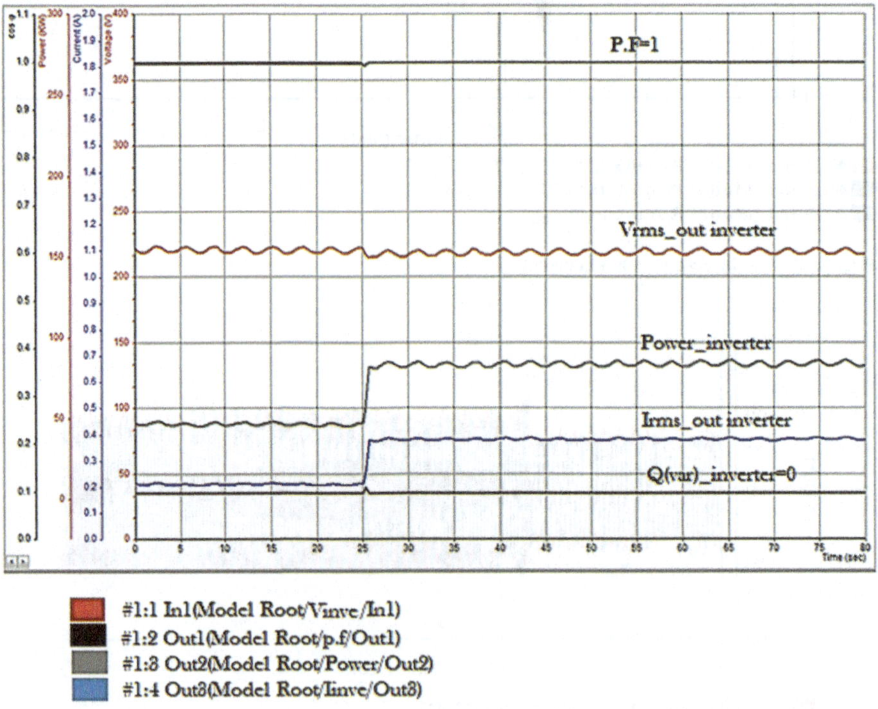

Fig. 6.31 Output inverter

Chapter 7
Conclusions and Future Work

7.1 Conclusions

The mathematical model of VSC-HVDC system is discussed in this book. Various abnormal conditions are simulated through the "MATLAB/Simulink" at both the grid terminals and the DC transmission line where it is connected to between two converters. Different control strategies are proposed for the VSC-based HVDC system. The behavior of different faults is studied under the different disturbances. The relay is used to protect the converter from grid current feeding during DC fault. In atrial to control the VSC-based HVDC system, active and reactive control under AC fault method is employed in this work to properly design the VSC control schemes. The following are the major outcomes of the study:

- The technology of VSC-HVDC transmission is developed and more utilized in renewable power energy applications.
- Control and system response performance under steady-state and fault conditions are tested and verified. Further, the design and control responses show an excellent matching.
- DC pole-to-pole fault has been selected to be analyzed because it is considered as one of the most dangerous faults in this system and may make damage to the system components.
- A complete VSC-HVDC transmission system has been simulated using MATLAB/Simulink, and the system has been tested in normal and fault conditions.
- The analysis and simulation show out the characteristics of DC and AC fault. The fault behavior has been studied starting from the fault that occurs to the system steady state.

N. F. Ibrahim, S. S. Dessouky, *Design and Implementation of Voltage Source Converters in HVDC Systems*, Power Systems,
https://doi.org/10.1007/978-3-030-51661-1_7

- It is concluded that in this type of faults, the system configuration changes in time, and it starts with RLC circuit, then transforms into a three-phase diode bridge, and ends with a current source supplying the fault.
- It is found from mathematical analysis that the DC fault passes through three stages from starting moment to steady-state moment:

  (a) Capacitor discharging stage
  (b) Diode freewheel stage
  (c) Grid feeding stage

- The converter output harmonic contents match the theoretical predictions from the PWM scheme. THD levels for both injected current and voltage at PCC are less than 2% by using LCL filter design.
- The rebalance of capacitors can be realized by using high impedance grounding system. In this recovery method, the discharging current is very small so that the voltage of the capacitor can maintain without any overcurrent stress.
- This book presents power control technique at grid-connected converters under unbalanced voltage conditions and short circuit faults. The current positive and negative sequences during faults are controlled to ensure a proper exchange of active and reactive powers without power ripples.
- System based on hysteresis current controller was used to generate the firing pulses of the VSI inverter, because it offers higher flexibility and simplicity in the experimental part.
- The results have shown that the supply voltage and current are almost in phase; good tracking of the current references has to be ensured.
- The total harmonic distortion of supply current is always small because it used LCL filter. The total harmonic distortion THD improved from 17.5% to 3.15%.

7.2 Suggestions for Future Work

The following points are recommended for future work in this field:

1. Analyze the effect of tuning of controller parameters on system performance and stability limits, taking into account the complexity of complete transfer function when there are several PI controllers in cascade.
2. Implementation of the developed control strategies in the $\alpha\beta$ stationary reference frame using PI controller; comparison of the results with the results obtained using dq synchronous reference frame.

Appendix

Appendix A: Datasheet of IGBT Modules

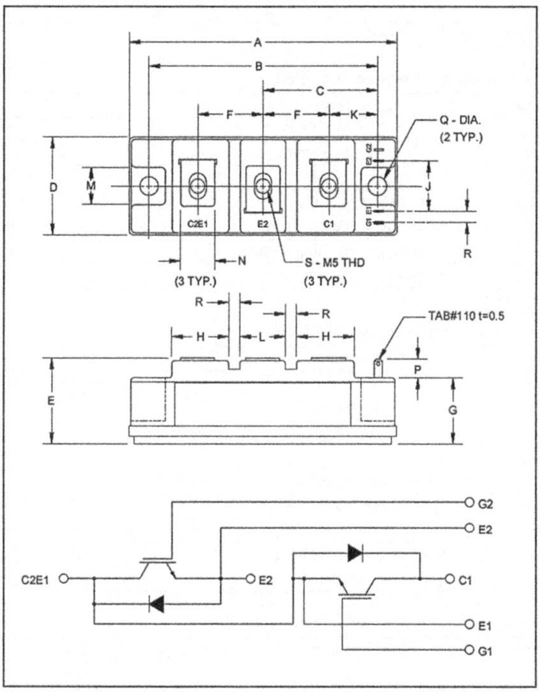

Outline Drawing and Circuit Diagram

Dimensions	Inches	Millimeters
A	3.70	94.0
B	3.150±0.01	80.0±0.25
C	1.57	40.0
D	1.34	34.0
E	1.22 Max.	31.0 Max.
F	0.90	23.0
G	0.85	21.5
H	0.79	20.0
J	0.71	18.0

Dimensions	Inches	Millimeters
K	0.67	17.0
L	0.63	16.0
M	0.51	13.0
N	0.47	12.0
P	0.28	7.0
Q	0.256 Dia.	Dia. 6.5
R	0.16	4.0
S	M5 Metric	M5

Description:
Mitsubishi IGBT Modules are designed for use in switching applications. Each module consists of two IGBTs in a half-bridge configuration with each transistor having a reverse-connected super-fast recovery free-wheel diode. All components and interconnects are isolated from the heat sinking baseplate, offering simplified system assembly and thermal management.

Features:
- ☐ Low Drive Power
- ☐ Low $V_{CE(sat)}$
- ☐ Discrete Super-Fast Recovery Free-Wheel Diode
- ☐ High Frequency Operation
- ☐ Isolated Baseplate for Easy Heat Sinking

Applications:
- ☐ AC Motor Control
- ☐ Motion/Servo Control
- ☐ UPS
- ☐ Welding Power Supplies

Ordering Information:
Example: Select the complete part module number you desire from the table below -i.e. CM50DY-24H is a 1200V (V_{CES}), 50 Ampere Dual IGBT Module.

Type	Current Rating Amperes	V_{CES} Volts (x 50)
CM	50	24

N. F. Ibrahim, S. S. Dessouky, *Design and Implementation of Voltage Source Converters in HVDC Systems*, Power Systems,
https://doi.org/10.1007/978-3-030-51661-1

Absolute Maximum Ratings, T_j = 25 °C unless otherwise specified

	Symbol	Ratings	Units
Junction Temperature	T_j	–40 to 150	°C
Storage Temperature	T_{stg}	–40 to 125	°C
Collector-Emitter Voltage (G-E SHORT)	V_{CES}	1200	Volts
Gate-Emitter Voltage (C-E SHORT)	V_{GES}	±20	Volts
Collector Current (T_C = 25°C)	I_C	50	Amperes
Peak Collector Current	I_{CM}	100*	Amperes
Emitter Current** (T_C = 25°C)	I_E	50	Amperes
Peak Emitter Current**	I_{EM}	100*	Amperes
Maximum Collector Dissipation (T_C = 25°C, $T_j \leq$ 150°C)	P_c	400	Watts
Mounting Torque, M5 Main Terminal	–	1.47 ~ 1.96	N · m
Mounting Torque, M6 Mounting	–	1.96 ~ 2.94	N · m
Weight	–	190	Grams
Isolation Voltage (Main Terminal to Baseplate, AC 1 min.)	V_{iso}	2500	Vrms

*Pulse width and repetition rate should be such that the device junction temperature (T_j) does not exceed $T_{j(max)}$ rating.
**Represents characteristics of the anti-parallel, emitter-to-collector free-wheel diode (FWDi).

Static Electrical Characteristics, T_j = 25 °C unless otherwise specified

Characteristics	Symbol	Test Conditions	Min.	Typ.	Max.	Units
Collector-Cutoff Current	I_{CES}	$V_{CE} = V_{CES}$, V_{GE} = 0V	–	–	1.0	mA
Gate Leakage Current	I_{GES}	$V_{GE} = V_{GES}$, V_{CE} = 0V	–	–	0.5	μA
Gate-Emitter Threshold Voltage	$V_{GE(th)}$	I_C = 5mA, V_{CE} = 10V	4.5	6.0	7.5	Volts
Collector-Emitter Saturation Voltage	$V_{CE(sat)}$	I_C = 50A, V_{GE} = 15V	–	2.5	3.4**	Volts
		I_C = 50A, V_{GE} = 15V, T_j = 150°C	–	2.25	–	Volts
Total Gate Charge	Q_G	V_{CC} = 600V, I_C = 50A, V_{GE} = 15V	–	250	–	nC
Emitter-Collector Voltage	V_{EC}	I_E = 50A, V_{GE} = 0V	–	–	3.5	Volts

** Pulse width and repetition rate should be such that device junction temperature rise is negligible.

Dynamic Electrical Characteristics, T_j = 25 °C unless otherwise specified

Characteristics		Symbol	Test Conditions	Min.	Typ.	Max.	Units
Input Capacitance		C_{ies}	V_{GE} = 0V, V_{CE} = 10V	–	–	10	nF
Output Capacitance		C_{oes}		–	–	3.5	nF
Reverse Transfer Capacitance		C_{res}		–	–	2	nF
Resistive	Turn-on Delay Time	$t_{d(on)}$	V_{CC} = 600V, I_C = 50A,	–	–	80	ns
Load	Rise Time	t_r	$V_{GE1} = V_{GE2}$ = 15V, R_G = 6.3Ω	–	–	200	ns
Switching	Turn-off Delay Time	$t_{d(off)}$		–	–	150	ns
Times	Fall Time	t_f		–	–	350	ns
Diode Reverse Recovery Time		t_{rr}	I_E = 50A, di_E/dt = –100A/μs	–	–	250	ns
Diode Reverse Recovery Charge		Q_{rr}	I_E = 50A, di_E/dt = –100A/μs	–	0.37	–	μC

Thermal and Mechanical Characteristics, T_j = 25 °C unless otherwise specified

Characteristics	Symbol	Test Conditions	Min.	Typ.	Max.	Units
Thermal Resistance, Junction to Case	$R_{th(j-c)}$	Per IGBT	–	–	0.31	°C/W
Thermal Resistance, Junction to Case	$R_{th(j-c)}$	Per FWDi	–	–	0.70	°C/W
Contact Thermal Resistance	$R_{th(c-f)}$	Per Module, Thermal Grease Applied	–	–	0.075	°C/W

Appendix B: Datasheet of Voltage Transducer

Voltage Transducer LV 25-P

I_{PN} = 10 mA
V_{PN} = 10 .. 500 V

For the electronic measurement of currents: DC, AC, pulsed..., with galvanic separation between the primary circuit and the secondary circuit.

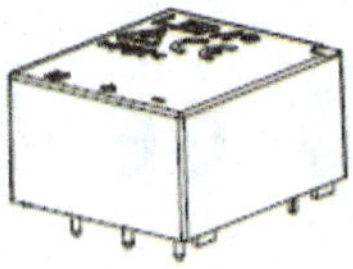

Electrical data

I_{PN}	Primary nominal rms current		10	mA
I_{PM}	Primary current, measuring range		0 .. ± 14	mA
R_M	Measuring resistance		$R_{M\,min}$ $R_{M\,max}$	
	with ± 12 V	@ ± 10 mA $_{max}$	30 190	Ω
		@ ± 14 mA $_{max}$	30 100	Ω
	with ± 15 V	@ ± 10 mA $_{max}$	100 350	Ω
		@ ± 14 mA $_{max}$	100 190	Ω
I_{SN}	Secondary nominal rms current		25	mA
K_N	Conversion ratio		2500 : 1000	
U_C	Supply voltage (± 5 %)		± 12 .. 15	V
I_C	Current consumption		10 (@ ± 15 V) + I_S	mA

Accuracy - Dynamic performance data

X_G	Overall accuracy @ I_{PN}, T_A = 25 °C @ ± 12 .. 15 V	± 0.9			%
	@ ± 15 V (± 5 %)	± 0.8			%
ε_L	Linearity error	< 0.2			%
		Typ	Max		
I_O	Offset current @ I_P = 0, T_A = 25 °C		± 0.15		mA
I_{OT}	Temperature variation of I_O 0 °C .. + 25 °C	± 0.06	± 0.25		mA
	+ 25 °C .. + 70 °C	± 0.10	± 0.35		mA
t_r	Step response time [1] to 90 % of I_{PN}	40			μs

General data

T_A	Ambient operating temperature	0 .. + 70	°C
T_S	Ambient storage temperature	- 25 .. + 85	°C
R_P	Resistance of primary winding @ T_A = 70 °C	250	Ω
R_S	Resistance of secondary winding @ T_A = 70 °C	110	Ω
m	Mass	22	g
	Standards	EN 50178: 1997 UL 508: 2010	

Note: [1] R_1 = 25 kΩ (L/R constant, produced by the resistance and inductance of the primary circuit).

Features

- Closed loop (compensated) current transducer using the Hall effect
- Insulating plastic case recognized according to UL 94-V0.

Principle of use

- For voltage measurements, a current proportional to the measured voltage must be passed through an external resistor R_1 which is selected by the user and installed in series with the primary circuit of the transducer.

Advantages

- Excellent accuracy
- Very good linearity
- Low thermal drift
- Low response time
- High bandwidth
- High immunity to external interference
- Low disturbance in common mode.

Applications

- AC variable speed drives and servo motor drives
- Static converters for DC motor drives
- Battery supplied applications
- Uninterruptible Power Supplies (UPS)
- Power supplies for welding applications.

Application domain

- Industrial.

Appendix C: Datasheet of Current Transducer

Current Transducer LA 25-NP

I_{PN} = 5-6-8-12-25 At

For the electronic measurement of currents: DC, AC, pulsed..., with galvanic isolation between the primary circuit and the secondary circuit.

16080

Electrical data

I_{PN}	Primary nominal current rms		25				At
I_{PM}	Primary current, measuring range		0 .. ± 36				At
R_M	Measuring resistance @		T_A = 70°C		T_A = 85°C		
			$R_{M\,min}$	$R_{M\,max}$	$R_{M\,min}$	$R_{M\,max}$	
	with ± 15 V	@ ± 25 At $_{max}$	100	320	100	315	Ω
		@ ± 36 At $_{max}$	100	190	100	185	Ω
I_{SN}	Secondary nominal current rms		25				mA
K_N	Conversion ratio		1-2-3-4-5 : 1000				
V_C	Supply voltage (± 5 %)		± 15				V
I_C	Current consumption		10 + I_S				mA

Accuracy - Dynamic performance data

X	Accuracy @ I_{PN}, T_A = 25°C		± 0.5		%
ε_L	Linearity error		< 0.2		%
			Typ	Max	
I_O	Offset current [1] @ I_P = 0, T_A = 25°C		± 0.05	± 0.15	mA
I_{OM}	Magnetic offset current [2] @ I_P = 0 and specified R_M, after an overload of 3 x I_{PN}		± 0.05	± 0.15	mA
I_{OT}	Temperature variation of I_O	0°C .. + 25°C	± 0.06	± 0.25	mA
		+ 25°C .. + 70°C	± 0.10	± 0.35	mA
		- 25°C .. + 85°C		± 0.5	mA
		- 40°C .. + 85°C		± 1.2	mA
t_r	Response time [3] to 90 % of I_{PN} step		< 1		μs
di/dt	di/dt accurately followed		> 50		A/μs
BW	Frequency bandwidth (- 1 dB)		DC .. 150		kHz

General data

T_A	Ambient operating temperature		- 40 .. + 85	°C
T_S	Ambient storage temperature		- 45 .. + 90	°C
R_P	Primary coil resistance per turn	@ T_A = 25°C	< 1.25	mΩ
R_S	Secondary coil resistance	@ T_A = 70°C	110	Ω
		@ T_A = 85°C	115	Ω
R_{IS}	Isolation resistance @ 500 V, T_A = 25°C		> 1500	MΩ
m	Mass		22	g
	Standards		EN 50178: 1997	

Features

- Closed loop (compensated) current transducer using the Hall effect
- Isolated plastic case recognized according to UL 94-V0.

Advantages

- Excellent accuracy
- Very good linearity
- Low temperature drift
- Optimized response time
- Wide frequency bandwidth
- No insertion losses
- High immunity to external interference
- Current overload capability.

Applications

- AC variable speed drives and servo motor drives
- Static converters for DC motor drives
- Battery supplied applications
- Uninterruptible Power Supplies (UPS)
- Switched Mode Power Supplies (SMPS)
- Power supplies for welding applications.

Application domain

- Industrial.

Dimensions LA 25-NP (in mm)

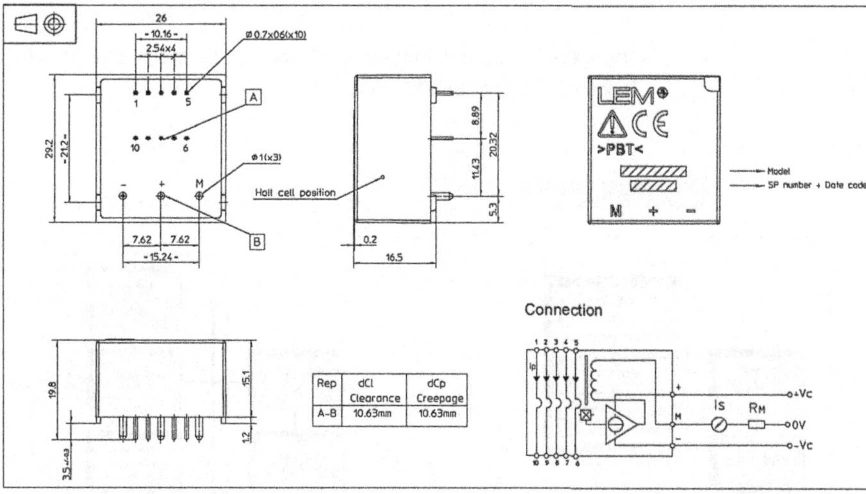

Number of primary turns	Primary current		Nominal output current I_{SN} [mA]	Turns ratio K_N	Primary resistance R_P [mΩ]	Primary insertion inductance L_P [μH]	Recommended connections
	nominal I_{PN} [A]	maximum I_P [A]					
1	25	36	25	1 / 1000	0.3	0.023	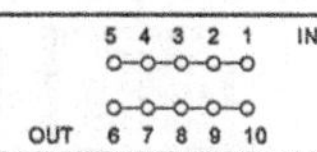
2	12	18	24	2 / 1000	1.1	0.09	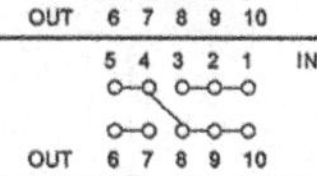
3	8	12	24	3 / 1000	2.5	0.21	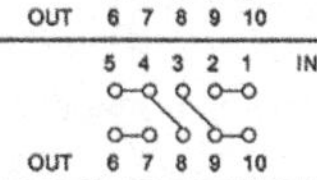
4	6	9	24	4 / 1000	4.4	0.37	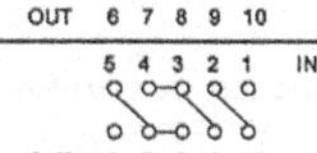
5	5	7	25	5 / 1000	6.3	0.58	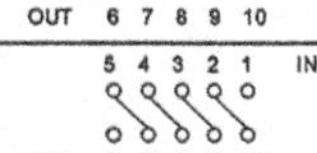

Mechanical characteristics

- General tolerance ± 0.2 mm
- Fastening & connection of primary 10 pins 0.7 x 0.6 mm
- Fastening & connection of secondary 3 pins Ø 1 mm
- Recommended PCB hole 1.2 mm

Remarks

- I_S is positive when I_P flows from terminals 1, 2, 3, 4, 5 to terminals 10, 9, 8, 7, 6.
- This is a standard model. For different versions (supply voltages, turns ratios, unidirectional measurements...),

Appendix D: Datasheet of DS1104

The following illustration gives an overview of the architecture and the functional units of the DS1104:

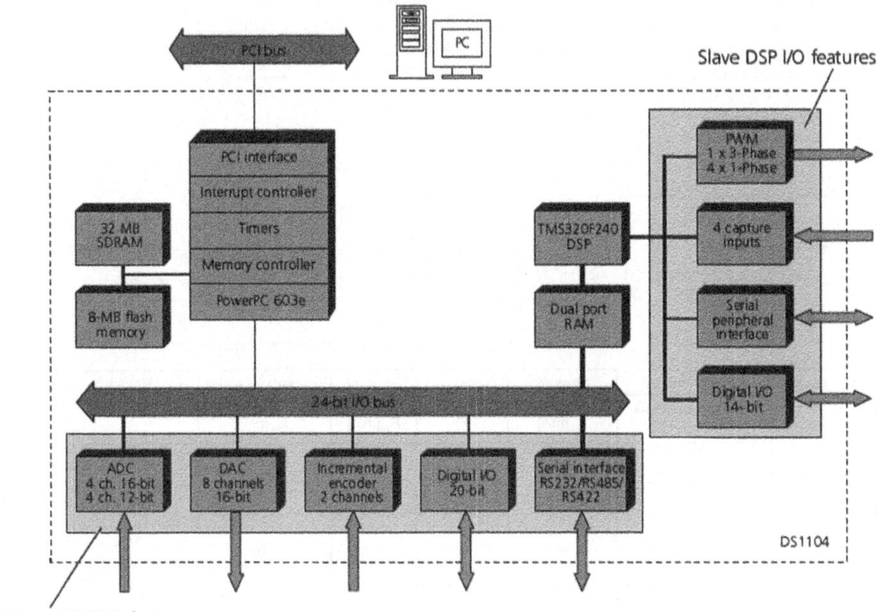

Feature overview

The DS1104 R&D Controller Board provides the following features:

Master PPC representing the computing power of the board, and featuring several I/O units (see *Features Provided by the Master PPC* on page 13)

Slave DSP featuring further I/O units (see *Features Provided by the Slave DSP* on page 37)

Interrupt controller providing various hardware and software interrupts (see *Interrupts Provided by the DS1104* on page 63)

Memory comprising DRAM and flash memory (see *Memory Features* on page 73)

Timers providing a sample rate timer, a time base counter, and 4 general-purpose timers (see *Timer Features* on page 75)

Appendix E: Datasheet of I/O Connector

I/O Connector (P1)	Pin	Sub-D Pin	Signal	Pin	Sub-D Pin	Signal
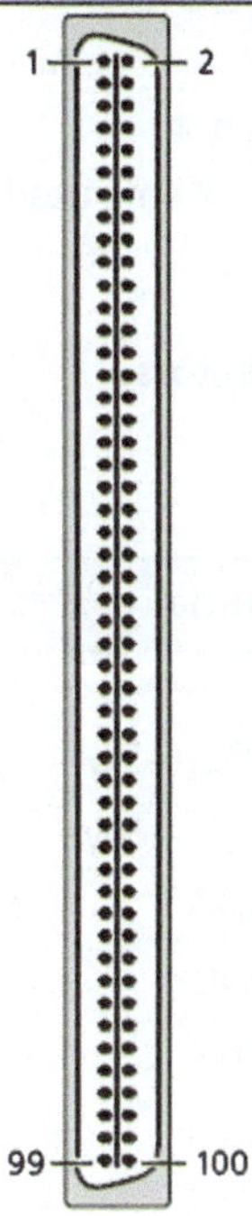	P1 1	P1B 1	GND	P1 2	P1A 1	GND
	P1 3	P1B 34	DCD (CTS)	P1 4	P1A 34	(RTS)
	P1 5	P1B 18	CTS ($\overline{\text{CTS}}$)	P1 6	P1A 18	RTS ($\overline{\text{RTS}}$)
	P1 7	P1B 2	DSR (RXD)	P1 8	P1A 2	DTR (TXD)
	P1 9	P1B 35	RXD ($\overline{\text{RXD}}$)	P1 10	P1A 35	TXD ($\overline{\text{TXD}}$)
	P1 11	P1B 19	SSOMI	P1 12	P1A 19	SCAP4
	P1 13	P1B 3	SSIMO	P1 14	P1A 3	SCAP3
	P1 15	P1B 36	SSTE	P1 16	P1A 36	SCAP2
	P1 17	P1B 20	SSCLK	P1 18	P1A 20	SCAP1
	P1 19	P1B 4	VCC (+5 V)	P1 20	P1A 4	VCC (+5 V)
	P1 21	P1B 37	ST3PWM	P1 22	P1A 37	SPWM6
	P1 23	P1B 21	ST2PWM	P1 24	P1A 21	SPWM5
	P1 25	P1B 5	ST1PWM	P1 26	P1A 5	SPWM4
	P1 27	P1B 38	SPWM9	P1 28	P1A 38	SPWM3
	P1 29	P1B 22	SPWM8	P1 30	P1A 22	SPWM2
	P1 31	P1B 6	SPWM7	P1 32	P1A 6	SPWM1
	P1 33	P1B 39	GND	P1 34	P1A 39	GND
	P1 35	P1B 23	$\overline{\text{IDX(2)}}$	P1 36	P1A 23	$\overline{\text{IDX(1)}}$
	P1 37	P1B 7	IDX(2)	P1 38	P1A 7	IDX(1)
	P1 39	P1B 40	$\overline{\text{PHI90(2)}}$	P1 40	P1A 40	$\overline{\text{PHI90(1)}}$
	P1 41	P1B 24	PHI90(2)	P1 42	P1A 24	PHI90(1)
	P1 43	P1B 8	$\overline{\text{PHI0(2)}}$	P1 44	P1A 8	$\overline{\text{PHI0(1)}}$
	P1 45	P1B 41	PHI0(2)	P1 46	P1A 41	PHI0(1)
	P1 47	P1B 25	GND	P1 48	P1A 25	GND
	P1 49	P1B 9	IO19	P1 50	P1A 9	IO18
	P1 51	P1B 42	IO17	P1 52	P1A 42	IO16
	P1 53	P1B 26	IO15	P1 54	P1A 26	IO14
	P1 55	P1B 10	IO13	P1 56	P1A 10	IO12
	P1 57	P1B 43	IO11	P1 58	P1A 43	IO10
	P1 59	P1B 27	IO9	P1 60	P1A 27	IO8
	P1 61	P1B 11	IO7	P1 62	P1A 11	IO6
	P1 63	P1B 44	IO5	P1 64	P1A 44	IO4
	P1 65	P1B 28	IO3	P1 66	P1A 28	IO2
	P1 67	P1B 12	IO1	P1 68	P1A 12	IO0
	P1 69	P1B 45	GND	P1 70	P1A 45	GND
	P1 71	P1B 29	DACH8	P1 72	P1A 29	DACH7
	P1 73	P1B 13	GND	P1 74	P1A 13	GND
	P1 75	P1B 46	DACH6	P1 76	P1A 46	DACH5
	P1 77	P1B 30	GND	P1 78	P1A 30	GND
	P1 79	P1B 14	DACH4	P1 80	P1A 14	DACH3
	P1 81	P1B 47	GND	P1 82	P1A 47	GND
	P1 83	P1B 31	DACH2	P1 84	P1A 31	DACH1
	P1 85	P1B 15	GND	P1 86	P1A 15	GND
	P1 87	P1B 48	ADCH8	P1 88	P1A 48	ADCH7
	P1 89	P1B 32	GND	P1 90	P1A 32	GND
	P1 91	P1B 16	ADCH6	P1 92	P1A 16	ADCH5
	P1 93	P1B 49	GND	P1 94	P1A 49	GND
	P1 95	P1B 33	ADCH4	P1 96	P1A 33	ADCH3
	P1 97	P1B 17	GND	P1 98	P1A 17	GND
	P1 99	P1B 50	ADCH2	P1 100	P1A 50	ADCH1

DS1104 Hardware Installation and Configuration March 2004

Digital I/O Connector (CP17)

The digital I/O connector (CP17) is a 37-pin, male Sub-D connector located on the front of the connector panel.

Pinout Because the pin numbering used for Sub-D connectors is not standardized, the following figure shows the numbering scheme used (front view).

Do not rely on the numbers written on Sub-D connectors.

Connector (CP17)	Pin	Signal	Pin	Signal
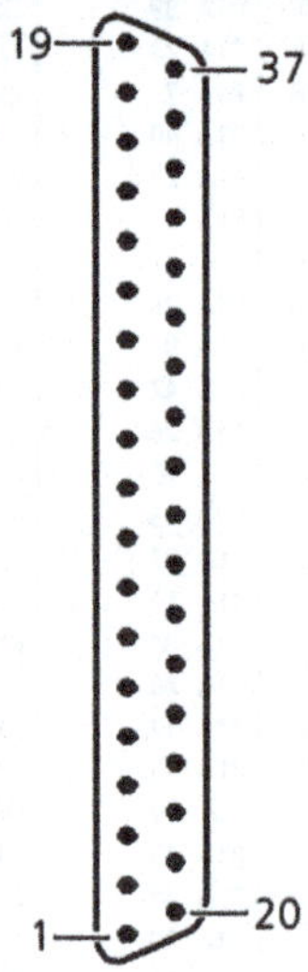	19	GND		
	18	GND	37	VCC (+5 V)
	17	GND	36	VCC (+5 V)
	16	GND	35	GND
	15	IO19	34	GND
	14	IO17	33	IO18
	13	GND	32	IO16
	12	IO15	31	GND
	11	IO13	30	IO14
	10	GND	29	IO12
	9	IO11	28	GND
	8	IO9	27	IO10
	7	GND	26	IO8
	6	IO7	25	GND
	5	IO5	24	IO6
	4	GND	23	IO4
	3	IO3	22	GND
	2	IO1	21	IO2
	1	GND	20	IO0

References

1. D.-I.S. Ana-Irina Stan, *Control of VSC-Based HVDC Transmission System for Offshore Wind Power Plants* (Department of Energy Technology Aalborg University, Denmark, 2010)
2. L. Zhang, Modeling and control of VSC-HVDC links connected to weak AC systems, Doctor of Philosophy, School of Electrical Engineering Electrical Machines and Power Electronics, Royal Institute of Technology, Stockholm, 2010
3. S. Chandhary, R. Teodorescu, P. Rodriguez, P. Kjaer, P. Christensen, Modelling and simulation of vsc-based hvdc connection for offshore wind power plants, Ph.D. Seminar on detailed Modelling and Validation of Electrical Components and Systems, pp. 53–57, 2010
4. H. Bewa, ABB develops complete system solution for 1,100 kV HVDC power transmission, 2014. [Online]. Available: http://www.abb.ch/cawp/seitp22/3a8302e9925218a4c1257d3f0451b52.aspx
5. R. L. Sellick, M. Akerberg, Comparison of HVDC light (VSC) and HVDC classic (LCC) site aspects, for a 500MW 400kV HVDC transmission scheme, in *10th IET International Conference on AC and DC Power Transmission, ACDC 2012*, Birmingham, UK, pp. 1–6, 2012
6. Dig SILENT Technical Documentation - PWM Converter, Dig SILENT GmbH, Tech. Rep., 2014
7. A. Fuchs, Coordinated control of power systems with HVDC links, PhD Thesis, ETH Zurich, 2015
8. A. Yazdani, R. Iravani, *Voltage-Sourced Converters in Power Systems: Modeling, Control, and Applications* (IEEE Press/Wiley, Hoboken, 2010)
9. N. Flourentzou, V.G. Agelidis, G.D. Demetriades, VSC-based HVDC power transmission systems: An overview. IEEE Trans. Power Electron. **24**(3), 592–602 (2009)
10. A. Kalair, N. Abas, N. Khan, Comparative study of HVAC and HVDC transmission systems. Renew. Sust. Energ. Rev. **59**, 1653–1675 (2016)
11. M. Davies, M. Dommaschk, J. Dorn, J. Lang, D. Retzmann, D. Soerangr, HVDC PLUS- Basics and principles of operation, Siemens AG-Energy Sector, 2011
12. B. Chuco, E.H. Watanabe, A comparative study of dynamic performance of HVDC system based on conventional VSC and MMC-VSC, in *Proceedings of Bulk Power System Dynamics and Control Symposium*, pp.1–6, 2010
13. M.A. Djehaf, S.A. Zidi, Y.I. Kobibi, S. Hadjeri, Modeling of a multi-level converter based VSC HVDC supplying a dead load, in *Proceedings of Electrical and Information Technologies International Conference*, pp. 218–223, 2015

N. F. Ibrahim, S. S. Dessouky, *Design and Implementation of Voltage Source Converters in HVDC Systems*, Power Systems,
https://doi.org/10.1007/978-3-030-51661-1

14. Siemens AG. ROBICON Perfect Harmony. [Online]. Available: http://www.industry.siemens.com/drives/global/en/converter/mvdrive/sinamicperfect-harmony/page/sinamics-perfct-harmony.aspx. Accesed Mar 2016
15. M. H. Okba, M. H. Saied, M. Z. Mostafa, T. M. Abdel-Moneim, High voltage direct current transmission- a review, part I, in *IEEE Energytech*, pp. 1–7, 2012
16. M. Bahrman, P.-E. Bjorklund, The new black start. System restoration with the help from voltage-source converters. IEEE Power and Energy Magazine **12**(1), 44–53 (2014)
17. G. Shilpa, P. Manohar, Hybrid HVDC system for multi-infeed applications, in *2013 International Conference on Emerging Trends in Communication, Control, Signal Processing & Computing Applications (C2SPCA)*, India, 10–11 October, pp. 1–5, 2013
18. V. Behravesh, N. Abbaspour, New comparison of HVDC and HVAC transmission system. Int. J. Eng. Innov. Res. **1**(3), 300–304 (2012)
19. O. Kotb, A hybrid HVDC transmission system supplying a passive load, M.Sc. Thesis, Department of Electrical and Computer Engineering, Oshawa, 2010
20. Power T&D Siemens, HVDC plus-basics and principle of operation, Book Siemens Energy Sector Power Transmission Solutions, 2008
21. S. Kouro et al., Recent advances and industrial applications of multilevel converters. IEEE Trans. Ind. Electron. **57**(8), 2553–2580 (Aug. 2010)
22. V. Manoj, K. Manohar, B.D. Prasad, Reduction of switching losses in VSC using DC link fuzzy logic controller innovative systems. Des. Eng. **3**(3) (2012)
23. P. Lundberg, M. Callavik, M. Bahrman, P. Sandeberg, High-voltage DC converters and cable technologies for offshore renewable integration and DC grid expansions. IEEE Power Energy Mag. **10**(6), 30–38 (2012)
24. L. Zhang et al., Interconnection of two very weak AC systems by VSC-HVDC links using power-synchronization control. IEEE Trans. Power Syst. **26**(1), 344–355 (2011)
25. Y. Li, Z.W. Zhang, C. Rehtanz, L.F. Luo, S. Ruberg, D.C. Yang, A new voltage source converter-HVDC transmission system based on an inductive filtering method. IET Gener. Transm. Distrib. **5**(5), 569–576 (May 2011)
26. T. Rastogi, M.T. Siddiqui, R. Sudha, K. Govardhan, Analysis of Thyristor based HVDC transmission system. Int. J. Electr. Eng. Technol. **3**(2), 29–38 (2012)
27. Y. Zhang et al., Voltage source converter in high voltage applications: Multilevel versus two-level converters, in *Proceedings of 9th IET International Conference on AC and DC Power Transmission*, London, U.K., pp. 1–5, 2010
28. G.P. Adam et al., Modular multilevel inverter: Pulse width modulation and capacitor balancing technique. IET Power Electron. **3**, 702–715 (2010)
29. G. P. Adam et al., Modular multilevel converter for medium-voltage applications, in *IEEE International Conference on Electrical Machines and Drives*, pp. 1013–1018, 2011
30. L. Xu, L. Yao, DC voltage control and power dispatch of a multi-terminal HVDC system for integrating large offshore wind farms. IET Renew. Power Gener. **5**(3), 223–233 (2011)
31. T. Haileselassie, Control, dynamics and operation of multi-terminal vsc-hvdc transmission systems, Ph.D. dissertation, Department of Electric Power Engineering, Norwegian University of Science and Technology (NTNU), Trondheim, 2012
32. G. Stamatiou, M. Bongiorno, Stability analysis of two terminal VSC-HVDC systems using the net-damping criterion. IEEE Trans. Power Delivery **31**(4), 1748–1756 (2016)
33. K.R. Padiyar, N. Prabhu, Modelling, control design and analysis of VSC based HVDC transmission systems, in *Power System Technology, 2014. Power Con 2014. 2014 International Conference*, Vol. 1, pp. 774–779, 2014
34. L. Shijie, L. Yaohua, A novel fast current-control method for the back-to-back converters, in *Industrial Technology, 2004. IEEE ICIT '04. 2004 IEEE International Conference*, Vol. 1, pp. 351–357, 2004
35. Z.S.A.K. Mohamed, H. Samir, F.M. Karim, A. Rabie, Performance analysis of a voltage source converter (VSC) based HVDC transmission system under faulted conditions. Leonardo J. Sci. (15), 33–46 (2009)
36. D. Cuiqing et al., A new control strategy of a VSC-HVDC system for high- quality supply of industrial plants. IEEE Trans. Power Delivery **22**, 2386–2394 (2007)

37. C. Du, The control of VSC-HVDC and its use for large industrial power systems, PhD, Department of Electric Power Engineering, Chalmers University of Technology, Goteborg, 2013
38. D. S. Cuiqing, A. Bollen, M. H. J, Analysis of the control algorithms of voltage-source converter HVDC, in *Power Tech, 2015 IEEE Russia*, pp. 1–7, 2015
39. J. Svensson, Grid-connected voltage source converter- control principles and wind energy applications, PhD thesis, Department of Electric Power Engineering, Chalmers University of Technology, Goteborg, 2011
40. T.M. Haileselassie, Control of multi-terminal VSC-HVDC systems, Master of Science in Energy and Environment, Department of Electrical Power Engineering, Norwegian University of Science and Technology, Trondheim, 2008
41. C. Bajracharia, Understanding of tuning techniques of converter controllers for VSCHVDC, Presented at the NORPIE, Finland, 2008
42. Y. Xue, Z. Xu, On the bipolar MMC-HVDC topology suitable for bulk power overhead line transmission: ConFiguration, control, and DC fault analysis. IEEE Trans. Power Delivery **29**, 2420–2429 (2014)
43. R. Li, G.P. Adam, D. Holliday, J.E. Fletcher, B.W. Williams, Hybrid cascaded modular multilevel converter with DC fault ride-through capability for the HVDC transmission system. IEEE Trans. Power Delivery **30**, 1853–1862 (2015)
44. S. Ademi, D. Tzelepis, A. Dysko, S. Subramanian, H. Ha, Fault current characterization in VSC-based HVDC systems, in *Proceedings of the IET Conference on Development in Power System Protection*, Edinburgh, UK, 7–10 March 2016
45. R. Li, L. Xu, D. Holliday, F. Page, S.J. Finney, B.W. Williams, Continuous operation of radial multiterminal HVDC systems under DC fault. IEEE Trans. Power Delivery **31**, 351–361 (2016)
46. Z. Suo, G. Li, R. Li, L. Xu, W. Wang, Y. Chi, W. Sun, Submodule configuration of HVDC-DC autotransformer considering DC fault. IET Power Electron. **9**, 2776–2785 (2016)
47. E.E. El-Kholy, A. El-Sabbe, A. El-Hefnawy, H.M. Mharous, Three-phase active power filter based on current controlled voltage source inverter. Int. J. Electr. Power Energy Syst., 537–547 (2006)
48. V. Anand, S.K. Srivastava, Simulation and performance investigation of series active power filter using hysteresis current control method. Int. J. Eng. Res. Appl. **2**, 1073–1080 (2012)
49. K. Ma, W. Chen, M. Liserre, F. Blaabjerg, Power controllability of a three-phase converter with an unbalanced AC source. IEEE Trans. Power Electron. **30**(3), 1591–1604 (2015)
50. J. Ferre, O.G. Bellmunt, T.C. Green, D.E.S. Sanchez, Current control reference calculation issues for the operation of renewable source grid interface vsc under unbalanced voltage sags. IEEE Trans. Power Electron. **26**(12), 3744–3753 (2011)
51. B. Yin, R. Oruganti, S.K. Panda, A.K.S. Bhat, An output-power-control strategy for a three-phase PWM rectifier under unbalanced supply conditions. IEEE Trans. Ind. Electron. **55**(5), 2140–2151 (2008)
52. N.F. Ibrahim, M. Fawzi, H.A. Ibrahim, S.S. Dessouky, Control strategies for VSC based HVDC during grid faults: A comparative study of selection criteria of currents reference. In 2019 21st International Middle East Power Systems Conference (MEPCON) (pp. 311–315). IEEE (2019).
53. M.A. Hannan, I. Hussin, P.J. Ker, M.M. Hoque, M.H. Lipu, A. Hussain, F. Blaabjerg, Advanced control strategies of VSC based HVDC transmission system. IEEE **44**(6), 78352–78369 (2018)
54. J. Khazaei, P. Idowu, A. Asrari, A.B. Shafaye, L. Piyasinghe, Review of HVDC control in weak AC grids. Electr. Power Syst. Res. **162**(5), 194–206 (2018)
55. T. Jing, A. S. A Maklakov, Review of voltage source converters for energy applications, in *IEEE International Ural Conference on Green Energy (UralCon)*, pp. 275–281, 2018
56. S.S. Dessouky, M. Fawzi, H.A. Ibrahim, N.F. Ibrahim, DC pole to pole short circuit fault analysis in VSCHVDC transmission system, in *2018 Twentieth International Middle East Power Systems Conference (MEPCON)*, Cairo University, Egypt, 2018

Index

N. F. Ibrahim, S. S. Dessouky, *Design and Implementation of Voltage Source Converters in HVDC Systems*, Power Systems,
https://doi.org/10.1007/978-3-030-51661-1

The manufacturer's authorised representative in the EU is Springer Nature Customer Service Centre GmbH, Europaplatz 3, 69115 Heidelberg, Germany. If you have any concerns regarding our products, please contact ProductSafety@springernature.com

Printed and bound by CPI Group (UK) Ltd, Croydon, CR0 4YY
28/09/2026
02224089-0016